Advanced Lignin Technologies

Edited by Antonio Pizzi

Published in London, United Kingdom

Advanced Lignin Technologies
http://dx.doi.org/10.5772/intechopen.1001669
Edited by Antonio Pizzi

Contributors
Akihiko Kondo, Antonio Pizzi, Belachew Zegale Tizazu, Chiaki Ogino, Filemon Jalu Nusantara Putra, Judith Emery Ngomoyogoli Kanemoto, Madurai Suguna Lakshmi, Olena Nedukha, Prihardi Kahar, Stanley Olivier Kanemoto, Tsegaye Abera Mekuriaw

First published in London, United Kingdom, 2024 by IntechOpen
IntechOpen is the global imprint of INTECHOPEN LIMITED, registered in England and Wales, registration number: 11086078, 167-169 Great Portland Street, London, W1W 5PF, United Kingdom

British Library Cataloguing-in-Publication Data
A catalogue record for this book is available from the British Library

Additional hard and PDF copies can be obtained from orders@intechopen.com

Advanced Lignin Technologies
Edited by Antonio Pizzi
p. cm.
Print ISBN 978-0-85466-764-2
Online ISBN 978-0-85466-763-5
eBook (PDF) ISBN 978-0-85466-765-9

For EU product safety concerns:
IN TECH d.o.o., Prolaz Marije Krucifikse Kozulić 3, 51000 Rijeka, Croatia,
info@intechopen.com or visit our website at intechopen.com.

Meet the editor

A. Pizzi is a professor emeritus of industrial chemistry at the University of Lorraine, DChem ([olymers, Rome, Italy), Ph.D. (organic chemistry, South Africa), and DSc (wood chemistry, South Africa). Experience gained in industry and research organizations allowed Dr. Pizzi to successfully serve as professor of polymer chemistry (1989–1995) and head of the Department of Chemistry (1991–1993) at the Witwatersrand University, Johannesburg, South Africa. In 1994, he moved back to Europe as a professor of industrial chemistry at LERMAB-ENSTIB, University of Lorraine, France. He has more than 1,000 research articles and patents to his name, and 10 books on bioadhesives and bioresins. He is the recipient of numerous international science prizes. His best area of specialization is in bioadhesives and bioresins chemistry, formulation, and application.

Contents

Preface

Biosourced materials research is all the fashion at present to build a green bioeconomy and green technology world. Lignin is the second most abundant biopolymer after cellulose, and unlike cellulose, it has not been extensively used in the manner that it should be. Several constraints have caused such a drawback, the main one of these being in principle the difficulty of unraveling its chemistry and adapting it to higher-value applications. Equally important is the initially perceived lower reactivity of such a material for a number of applications to which its structure seemed to be well suited, for example, to substitute phenol in phenol-formaldehyde resins. Some of these have caused a mountain of literature, unfortunately without leading to much worthy application. It is when more fundamental work and in particular very different approaches not anchored on old preconceived ideas based on petrochemicals have been taken that this field has become unblocked. It is a very interesting field not only for the abundance of the raw material but even more for its rather low relative costs making it economically very interesting.

The book presented here is divided into two main sections:

Section 1, entitled "Fundamentals", includes two chapters:

1. "Influence of Environmental Change on Monolignols and the Micromorphology of Leaf Epidermis in Hydrophytes and Terrestrial Plants" by Olena Nedukha

This chapter is very unusual and fundamental, tracing what really are the structure, morphology, and characteristics of lignin under different conditions and under different types of stress. This review also discusses data and results of experimental studies on the cellular mechanisms of lignin synthesis and its regulation, as well as the participation of monolignols in plant adaptation to environmental changes. It is a real must-read. Even specialists in lignin chemistry will be amazed at the science presented.

2. "Deep Eutectic Solvents for Lignin Isolation" by Tsegaye Abera Mekuriaw and Belachew Zegale Tizazu

This chapter takes an in-depth review of the recent study on deep eutectic solvents (DES), including their classifications, preparation methods, physicochemical characteristics, factors affecting DES pretreatment, and their applications. Based on the type of DES, molar ratio, type of biomass, reaction time, and reaction temperature, DESs can modify lignin characteristics.

Section 2, entitled "Advanced Applications", includes three chapters:

1. "Advanced Lignin Valorization for Biorefinery Application" by Filemon Jalu Nusantara Putra et al.

This chapter explores the immense potential of lignin biorefining, examining emerging biochemical strategies, including thermochemical and enzymatic depolymerization and physical techniques which have emerged as promising avenues for converting lignin into valuable biofuels and chemical compounds. By highlighting innovative approaches and technologies, it emphasizes the pivotal role of lignin in driving biorefineries toward low-emission processes, yielding a diverse spectrum of bioproducts.

2. "Low Ecological Impact Lignin-Based Flame Retardants for Polyurethane Foams" by Stanley Olivier Kanemoto et al.

In response to growing environmental concerns, lignin is an abundant and renewable resource, which has been employed to develop effective flame-retardant polyurethanes, with a simultaneous focus on reducing their ecological impact. Diverse chemical modifications of lignin have been explored to further enhance its fire resistance. This review highlights relevant advancements in this field.

3. "Recent Developments in Advanced Lignin-Based Adhesives and Binders" by Antonio Pizzi

Recently, several developments in the use of lignin and its derivatives as the wood adhesive and for other binders have come to the fore. This review chapter is focused on the types of more recent alternative approaches that have been used with encouraging results to go around the obstacle of the low lignin–aldehyde reactivity. Thus, approaches based on lignin demethylation coupled with specific oxidation, pre-glyoxalated lignin dialdehyde starch cross-linking by urea, lignin-based non-isocyanate polyurethane (NIPU) adhesives and resins, lignin amine coatings, lignin-triethyl phosphate cross-linking for both wood surface coatings and bio binders for metals/Teflon assemblies, and finally direct wood bonding by lignin esterification by citric acid are described.

Antonio Pizzi
LERMAB-ENSTIB,
University of Lorraine,
Epinal, France

Section 1

Fundamentals

Chapter 1

Influence of Environmental Change on Monolignols and the Micromorphology of Leaf Epidermis in Hydrophytes and Terrestrial Plants

Olena Nedukha

Abstract

The review presents data on the role of leaf epidermis, lignin, and its monolignols in the adaptation of higher plants to adverse environmental conditions. It discusses the localization and content of syringyl monolignin and guajacyl monolignin in cell walls and how they affect the elasticity of plant cell walls under different natural conditions. These data are presented using modern methods of laser scanning confocal microscopy, and scanning electron microscopy. The review also discusses literature data and the results of our own experimental studies on the cellular mechanisms of lignin synthesis and its regulation, as well as the participation of monolignols in plant adaptation to environmental changes. It shows the role of monolignols in regulating water balance and transpiration of plants, as well as in protecting plant cells from abiotic stresses and environmental changes. Recent studies have also shown the influence of lignin on the expression of genes involved in the synthesis of secondary cell walls and metabolites with protective properties. This review indicates the potential for further research into the role of monolignols and confirms that the conservation of species characterized by increased lignin synthesis may be a conceptual basis for the protection and conservation of flora from abiotic stresses.

Keywords: cell wall structure, syringyl, guajacyl, flooding, drought, cold, stress, genes

1. Introduction

Over the past decade, research has shown that the structure and composition of plant epidermis cell walls change in response to various environmental factors. This applies to both terrestrial plants and hydrophytes, which grow in water or on the shore near water. Environmental changes can disrupt cell functions such as growth, water and sun photon absorption, transpiration, strength, and protection. The epidermis of leaves and stems serves as the first barrier between the plant and its environment, protecting it from biotic and abiotic factors such as flooding, extreme temperatures, and pathogen invasion [1–3]. The leaf epidermis is defined by a combination of structural characteristics and the chemical composition of cell walls, which can change under stress conditions. Trichomes and epidermal wax in the epidermal cells play a role in regulating the absorption of photon energy, decreasing photoinhibition, and reflecting ultraviolet rays. Additionally, the cells of the leaf epidermis play a key role in the development of plants [4–6]. It is known that one of the main components of the epidermal cell walls, responsible for mechanical strength, leaf elasticity, and water and aqueous solution transport, is lignin and its constituent monolignols. These properties of lignin depend on the content and ratio of its constituent monolignols, which are formed in independent ways [7].

Drought and soil flooding are factors that impact plant growth and species conservation. The study of these environmental factors began with the development of nature ecology and has led to new possibilities for investigating the effects of environmental stress on plant species conservation and vital plant functions. It has been established that the biogenesis and function of leaf epidermis are mediated by the activation of specific enzymes in plants. These studies are described in a series of reviews [7–9]. Understanding the structural and functional characteristics of leaf epidermis under normal conditions and adapting to stress factors is important. This chapter reviews current knowledge about the structure and function of the epidermis of hydrophytes and terrestrial plants, as well as the participation of lignin and its monolignols under the influence of environmental change.

2. The impact of flooding on leaf epidermis and monolignols

Flooding is a potentially harmful stress for land plants caused by both natural (river flooding, heavy precipitation, and sea tides) and artificial causes (construction of reservoirs and ponds). It can be short-term or long-term, leading to the death of some plant species and crops. Flooding occurs when soils are waterlogged or when irrigated lands are not properly maintained. The combination of changes in salinity and impaired oxygen respiration in the root system inhibits aerobic processes, impairs the absorption of ions from the soil, and affects growth processes. Aerenchyma is formed in the roots, aerobic processes are inhibited, and fermentation begins with the formation of lactic acid, leading to acidification of the environment around the root system [10]. When flooded, plants are affected by reduced lighting, changes in spectrum, a lack of oxygen, and CO_2. Water absorbs light flux, and the attenuation is directly proportional to the water depth. Light is also absorbed by organic and inorganic particles in the water, with the long-wavelength parts of the spectrum being absorbed at greater depths [11].

Underwater plant leaves are impacted by a lack of light and CO_2, which has high solubility but a low diffusion coefficient in fresh water (1.7×10^{-9} m^2/c at 20°C). The pH of fresh water also affects gas solubility and the ratio of CO_2 to O_2 in water [12]. The air contains 0.03% carbon dioxide, 78.09% nitrogen, and 20.95% oxygen. In water, the amount of dissolved oxygen depends on atmospheric pressure, with higher pressure leading to greater solubility of oxygen. At 1 atm and 25°C, 0.023 g of oxygen is dissolved in 1 kg of water. The solubility of carbon dioxide in water is 200 times higher than that of oxygen [11]. The CO_2 content in fresh water is 350 mM m^{-3}, and the diffusion of gases in water is also lower. Underwater plants require 30 times more free CO_2 to saturate their leaves [12] and therefore use CO_2 ions as a carbon source [13]. Growth conditions in aquatic environments can cause hydrophytes to die when light is limited, and oxygen levels in the water and in plants decrease, leading to the inhibition of carbohydrate metabolism [14].

These adaptations allow hydrophytes to efficiently carry out photosynthesis and obtain the nutrients they need despite the challenging conditions. Additionally, some hydrophytes have specialized air channels and aerenchyma tissue that facilitate gas exchange and help them survive in waterlogged environments. These adaptations are essential for the survival of hydrophytes and marsh plants in their unique habitats [2, 15].

The features aid in acclimatization and support plant resistance to underwater existence, allowing for CO_2 to enter photosynthetic cells through diffusion in the outer epidermal walls. The structure of cell walls in underwater leaves and stems differs from that of surface hydrophytes and mesophyte leaves, highlighting the importance of comparative structural and functional studies in understanding plant adaptation to aquatic environments.

2.1 The submerged leaf epidermis structure

The outer cell walls of the epidermis and leaf stomata act as the first barrier and transport route for CO_2 and water, as well as the point of contact with the environment for plant organs. The cuticle of mesophyte leaves' epidermal walls contains pores that facilitate cuticular transpiration and water absorption. The functionality of cuticular pores in aerial leaves is influenced by the temperature and humidity of the surrounding air [16, 17]. In flooded plants, the cuticle is also synthesized in the epidermis and deposited in the periclinal cuticle sheaths. In certain species of higher aquatic plants (*Menta aquatica, Oenanthe aquatica*, and *Rumex palustris*) that are tolerant to complete flooding, old aerial leaves often die off underwater, giving way to new dissected leaves with a different cell structure [2, 18, 19] and altered cuticle structure [20, 21], up to and including its hydrolysis [22].

Aquatic plants, also known as hydrophytes, have a super hydrophobic cuticle on the epidermis of their leaves and stems. This prevents the formation of a water film on the surface of submerged organs, which leads to normal gas exchange in the leaves and to photosynthesis [23, 24]. Recent studies have focused on the structure of the cuticle of epidermal cells in underwater plants, particularly. The presence of cuticular pores in the cell walls of the epidermis of underwater leaves and the functioning of the epidermal cell wall were studied [15, 25, 26].

The study of the ultrastructural peculiarities of wall epidermis in submerged leaves of *Myriophyllum spicatum, Potamogeton perfoliatus*, and *Potamogeton pectinatus* water plants showed the presence of a very thin cuticle layer with pores of various sizes and densities (**Figure 1**) [25, 27].

The surface of the leaf epidermis of *M. spicatum* was covered with a continuous, uneven layer of cuticle, and individual epidermal cells were not distinguishable; chloroplasts were in epidermis cells (**Figure 1b**). In the epidermis of *M. spicatum,* cuticular pores were in a wavy, round, or straight form (**Figure 1c** and **d**). Their length was up to 130 nm, the diameter of the pore was 20 ± 4 nm, and the density of pores was 12–15 per 1 mm of cuticle extension. The epidermis of *P. pectinatus* leaves covered with almost parallel cuticle ridges (**Figure 1e–g′**), the height of which varied from 1.5 to 5 μm and the length from 12 to 45 μm. Due to the continuous cuticle layer, the boundaries of individual cells were not distinguishable. In contrast, the epidermal cells of *P. perfoliatus* leaves (**Figure 1b**) had clear walls. The walls of the epidermis in *P. perfoliatus* were two times higher in the wall and five to seven times the thickness of the periplasmic layer, and the thickness of the cuticle decreased. The cuticle pores

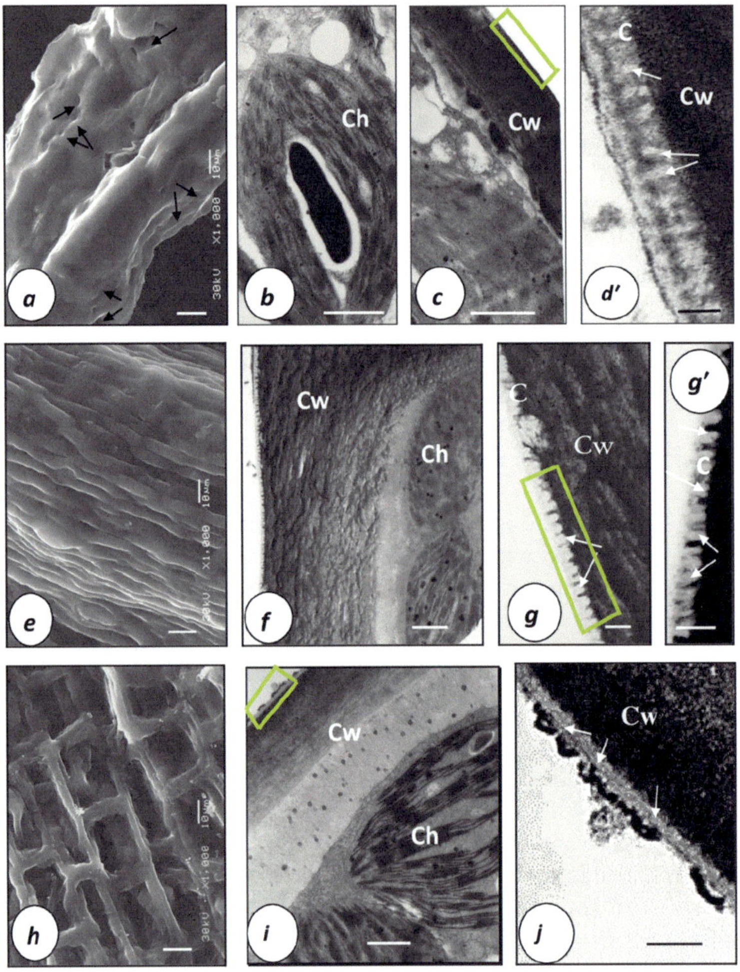

Figure 1.
Scanning electron microscopy of the epidermal cells of the submersed leaves of Myriophyllum spicatum (a, b, c, d), Potamogeton pectinatus (e, f, g, g'), and P. perfoliatus (h, I, j). Cuticular pores are indicated by arrows in figures a, d, g, g', and j. Abbreviations: C, cuticle; Cw, cell wall; Ch, chloroplast. Scale bars = 10 μm (a, e, h); = 5 μm; = 1 μm (b, c, f, i); = 100 nm (d, g, g', j), [25].

were faintly marked; the average size of the pores was about 4–5 nm. In contrast, the outer wall of the epidermis of *P. pectinatus* was very wide (up to 5 μm). Cuticle pores were located in the outer layer of the wall (**Figure 1f** and **j**); the average diameter of the pore was 17 ± 3 nm, and the length was 80–100 nm. The density of pores was 20 per 1 μm of cuticle extension. The authors suggest that the presence of cuticle pores in the walls of the epidermis of *M. spicatum*, *P. perfoliatus*, and *P. pectinatus* is a morphological sign of the adaptation of these species to submerged conditions. It is known that a very thick cuticle is characteristic only of mesophytes and xerophytes, and the cuticle pores of walls are very narrow [17]. The functioning of cuticle pores depends on endogenous and exogenous factors, including leaf age, air moisture, and air temperature [28]. The cuticle of the outer cell wall of the submerged leaf epidermis is involved in the transport of CO_2 and HCO_3 from water into the apoplast, then to the protoplast for photosynthesis [2].

2.2 Monolignols in submergenced leaves of hydrophytes

This section will discuss the presence of monolignols in the leaves of hydrophytes that are subjected to flooding. Lignin is essential for plant adaptation to stress, especially in response to flooding and siltation. It is a polymer of aromatic alcohols synthesized in cell walls that have completed stretching growth. Lignin helps make cell walls impermeable to water and aqueous solutions, forming a barrier to pathogens in the epidermis. It is composed of monolignins formed from p-hydroxyphenyl, guajacyl, and syringyl components, which differ in the degree of methoxylation [29]. Flooding and siltation impact the lignification of cell walls, with flooding being the main effect and mechanism of adaptation of the rice root system due to a lack of oxygen. The study investigated the impact of 50 cm deep flooding on the lignification of rice stems by comparing three rice varieties: flood-resistant (*FR13A*), flood-sensitive (*IR42*), and flood-adapted (Sabita). The flood-resistant mutant, *FR13A*, carries the genes *SUB1A* and *SUB1B* on the ninth chromosome, which encode transcription factors responsible for ethylene synthesis [30]. When subjected to flooding, the flood-sensitive variety *IR42* showed a significant increase in stem growth rate and leaf tensile strength compared to the flood-resistant variety. A negative correlation was found between stem elongation, lignin content, and the activities of coniferol alcohol dehydrogenase, and phenylalanine ammonium lyase. Additionally, a negative relationship between plant height and the content of structural polysaccharides (cellulose and hemicelluloses) was also observed. This suggests that flood-sensitive rice plants may have a different mechanism for coping with flooding stress compared to flood-resistant varieties.

According to researchers [31, 32], underwater (submerged) plant organs experience stress from the weight of the water column and the mechanical action of waves, leading to tension in flooded organs. The effect of this stress on the structure and functioning of cell walls in flooded plants are still not fully understood. While there is data on the impact of stress on the cell walls of land plants, the specific effects on flooded plants are not well documented. For example, in naked-seeded plants, the wood stress response occurs at the level of several vessel elements that have a small diameter, thick cell walls, high cellulose content, and low lignin [33]. These walls are characterized by a small amount of guajacol, a high content of syringin, and a three-layer structure [34].

A study examined the presence of lignin and monolignols in the leaves of the hydrophytes *Myriophyllum spicatum, Potamogeton pectinatus,* and *P. perfoliatus* (**Figure 2**) [35, 36]. Cytochemical analysis of monolignols in different tissues of submerged leaves of these three species revealed a higher content of syringyl and guajacyl in the epidermis cell walls compared to the walls of the photosynthesizing parenchyma. The study showed that the cell walls of underwater leaves of *M. spicatum* contained syringyl and guajacyl components, which emitted blue and green fluorescence, respectively, when combined with diphenyl boric acid-2-aminoethyl ester (DPBA) (**Figure 2b–d**). High levels of both guajacyl and syringyl were found in the cell walls of vessels and the anticlinal walls of the epidermis, while in the cell walls of the parenchyma, these levels were almost half as low. It is worth noting that the fluorescence of monolignins was significantly higher in the corners of the parenchymal cells compared to along the walls. Additionally, the ratio of syringyl to guajacyl content in the walls of photosynthetic parenchyma was quite high, especially in the first layer of photosynthetic parenchyma. In terms of the S/G ratio in leaf *M. spicatum*, cells are arranged in the following order: epidermis > vessels > parenchyma. Whereas

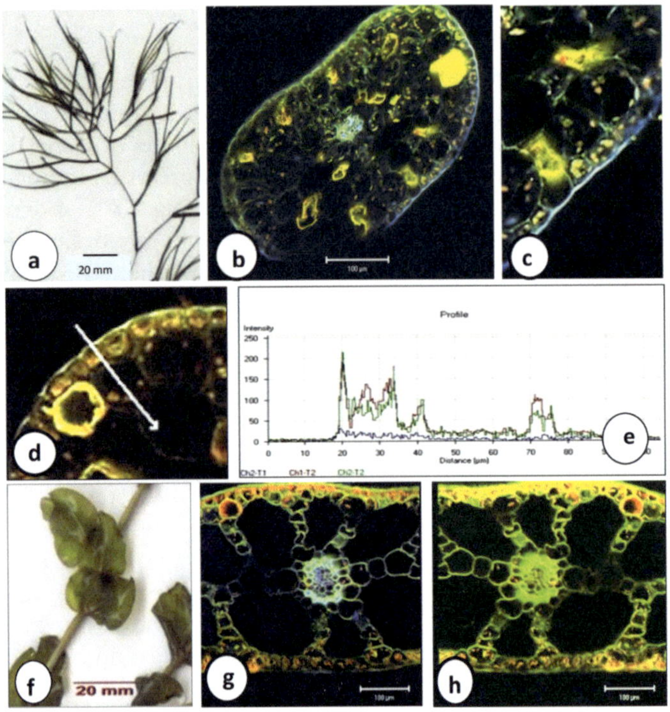

Figure 2.
The general view of underwater leaves of Myriophyllum spicatum and Potamogeton perfoliatus and the cytochemical fluorescence of monolignols bound to specific indicators diphenyl boric acid-2-aminoethyl ester (DPBA) in the leaves of M. spicatum leaves and P. perfoliatus was visualized with confocal laser scanning microscopy. Syringyl has blue fluorescence; guajacyl has green fluorescence; chlorophyll has red fluorescence. The histogram in figure e shows the fluorescence intensity of guajacyl, syringyl, and chlorophyll autofluorescence. The abscissa represents the distance (μm) scanned in figure d, which is shown by the white line. Scale bars: 100 μm [35].

in other species—*P. pectinatus*, this S/G ratio was in different order: parenchyma > epidermis > vessels > walls in the corners of parenchyma [36]. The high levels of monolignins in the corners of epidermal cells suggest an increase in the strength of these cells, making it difficult for pathogens to invade [37]. This epidermis also act as a barrier and protection of the leaf from the action of ultraviolet, which provokes the intensified lignin synthesis [38].

The presence and distribution of lignin in the tissues of underwater and above-water leaves of the hydrophyte *Sagittaria sagittifolia,* as well as the increased lignin content in submerged leaves, are similar to the data on the increase in lignin in flooded stems of *Ludviga repens* [39]. Little demonstrated that flooding *L. repens* stems increases the lignin content by 1.6 times compared to stems that are not flooded [39]. And this is what helps the plant to keep the flooded stem upright. Submerged leaves lack trichomes, stomata, and thick cuticles [15], making the surface of plant organs submerged in water more sensitive to pathogen invasion. Leaves and stems synthesize lignin in response to pathogen attack [37]. Exogenous and endogenous factors cause increased lignin synthesis in the underwater leaves of the studied species of water nut and arrowroot. Submerged plant organs must withstand water pressure and wave action, which affect the growth and structural and functional parameters of tissues, cells, and their walls [32, 40]. Lignification in the cell walls of parenchymal cells and vessels in mesophytes begins in the corners of cells and in the middle plate, which is filled with pectin substances. These zones become enriched with hydroxyphenyl and guajacol units [32, 41], as well as phenols such as coumaric and ferulic acids, which act as a foundation for lignin synthesis [42, 43].

2.3 Floating and above-water leaves

The epidermal structure of floating leaves is adapted to their aquatic environment. These leaves have a thick waxy cuticle and a reduced number of stomata, which helps to minimize water loss. Additionally, the epidermal cells may contain air spaces to provide buoyancy.

The micromorphology and ultrastructure of the epidermis of floating leaves of *Hydrocharis morsus-ranae* showed some differences from those of submerged leaves. The floating leaves of *H. morsus-ranae* are hypostomal, with differences in the ultrastructure and size of the main cells of the upper and lower epidermis. The upper epidermis cells are polygonal, oval, occasionally rounded, and tightly adjacent to each other (**Figure 3a**).

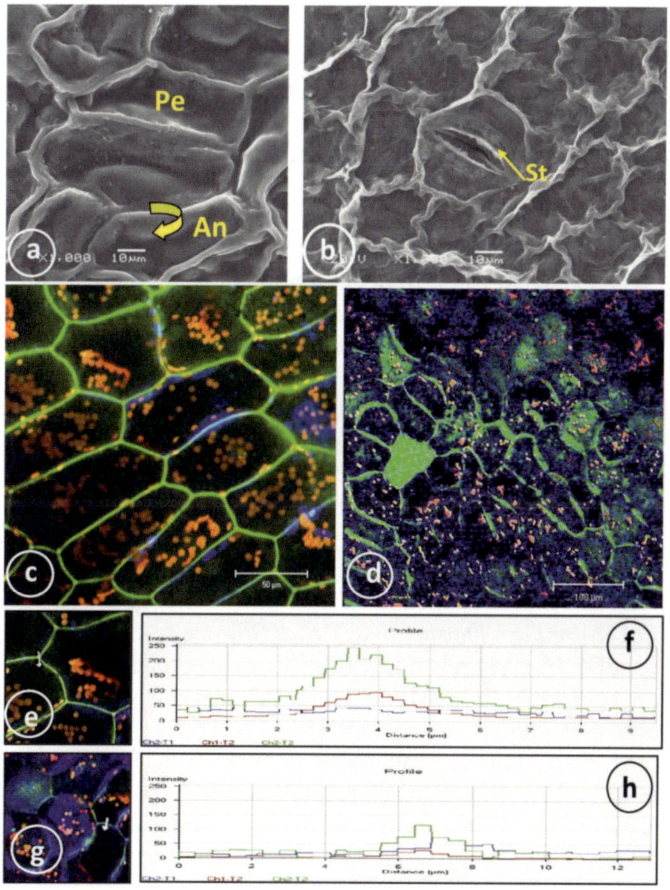

Figure 3.
The scanning electron microscopy images display the adaxial (a) and abaxial (b) surfaces of Hydrocharis morsus-ranae leaves. The fluorescence of syringyl and guajacyl is visualized with laser confocal microscopy on the adaxial (c and e) and abaxial (d, g) surfaces of the leaves and it is also shown on histograms (f and h). Abbreviations: An—anticlinal wall, pe—periclinal wall, and St—stomata. Scale bars: A, and b = 10 μm, c = 50 μm, and d = 100 μm.

The anticlinal walls are elevated above the periclinal walls, forming a cup-shaped depression in each cell. Stomata are located on the lower epidermis in no particular order (**Figure 3b**), with an average density of 33.1 ± 1.7 per 1 μm^2 of area. They are of the paracytic type, with accompanying cells located along the guard cells. The usual epidermal cells of the lower epidermis, similarly to those of the upper epidermis, form cushion-like depressions. The epidermal cells vary in shape, with round cells having a diameter of about 10 ± 1.2 μm and oval cells having a long axis ranging from 10 ± 0.7 to 37 ± 2.3 μm and a short axis from 7 ± 0.3 to 22 ± 1.9 μm, respectively. The anticlinal walls are wavy, reaching a height of 7 ± 0.4 μm. Periclinal walls are concave within the leaf blade and are located in depressions. Using the Pascal program (LSM), it has been calculated that there is air under the lower surface of the leaf, in the recesses of the main epidermal cells, with a content of 119,700 ± 2300 μm^3 on an area of 1 cm^2. Aerenchyma was also revealed between cells of the mesophyll and epidermis. It was established that the leaves of *H. morsus-ranae have* slightly deeper stomata in the epidermis, which is situated on the lower leaf surface. The presence of stomata on the lower epidermis of *H. morsus-ranae* was first described by Ancibor [44]. This investigator noted stomata deepened placement in the epidermis to increase the buoyancy of the leaves and not only for gas exchange. The author writes only words about the complete absence of lignification of the leaves of the toadstool or its very low content (although this author's figures are not shown), which ensures the flexibility of organs in water.

Our investigation found that the air content under the lower surface of the flooding leaf of *H. morsus-ranae* was 119,700 ± 2300 μm^3 per 1 cm^2. We also observed differences in the height of the anticlinal walls on the upper and lower surfaces of the leaves. The height of the anticlinal wall in the upper epidermis was one and a half times larger than that of the cells of the lower epidermis, and its height was 7 ± 0.4 μm. Similar protrusions of anticlinal walls were found in the surface and submerged leaves of *S. sagittifolia* [45]. Leaves of free-floating hydrophyte species, such as *Anubias barteri, Hydrophila odora, Bacopa caroliniana*, and *Hydrocharis* sp., characterized by presence of aerenchyma in leaf petioles and pedicels [46], and also by poorly developed mechanical and conductive tissue in leaf [47]. Aerenchyma was also detected in the stems of species collected in South Korea [48] and in the *Nelumbo nucifera peltate* leaf [49].

We demonstrated that syringyl and guajacyl monolignols, when combined with the specific fluorescent indicator diphenyl boric acid-2-aminoethyl ester, exhibited bright blue and green fluorescence in the periclinal and anticlinal walls of adaxial and abaxial epidermal cells of *Hydrocharis morsus-ranae* leaves (**Figure 3c–h**). It was shown that the intensity profile of DPBA-syringyl and DPBA-guajacyl complexes varied depending on the surface of the epidermis, the cell type. The anticlinal walls of usual cells in both epidermis, and the walls of stomata are characterized by highest fluorescence intensity. The intensity profile of syringyl in the periclinal and anticlinal walls of usual cells of adaxial surface was 27 ± 1.1 and 39 ± 1.3 relative units, respectively. On the abaxial surface, it was 91 ± 7.2 and 190 ± 11 relative units, respectively. The intensity profile of guajacyl was higher than that of syringyl in walls, measuring 53 ± 4.1 in the periclinal wall and 118 ± 13 relative units in the anticlinal wall of usual cells in the adaxial epidermis. The intensity profile of guaiacyl in such walls of the abaxial epidermis was significant in stomata, periclinal, and anticlinal walls, measuring 100 ± 10 and 120 ± 11 relative units, respectively, and also in the anticlinal wall of typical cells, measuring 91 ± 7.1 relative units. The intensity of guaiacyl in the periclinal wall of typical epidermal cells did not differ from that on the adaxial surface, measuring 115 ± 10.1 units.

Previous research has shown that an increase in the quantity of syringyl/guajacyl (S/G) acts as a chemical barrier, enhancing cell defense against water penetration and pathogens [50] while also increasing the mechanical durability of cells [51]. Based on our findings, we hypothesize that the elevated content of syringyl in the abaxial epidermis of the studied species can enhance the mechanical properties of the cell walls. Interestingly, the content of guajacyl was found to be similar in the upper and lower epidermal walls. Given the known function of guajacyl in wall softening, it is plausible that its high content in the cell walls of floating leaves of *H. morsus-ranae* promotes optimal movement of the leaves and the plant itself under the influence of water current waves.

S. sagittifolia and *Trapa natans* are convenient for studying the structure of above-water leaves of hydrophytes [26, 36, 45, 52, 53]. The ultrastructure of epidermal and mesophyll cells of the leaves of *S. sagittifolia shows* differences in cell wall structure between above-water and submerged leaves (see **Figure 4a–d**). The outer walls of the adaxial epidermis cells of above-water leaves of *S. sagittifolia* were about 1.5 μm wide and

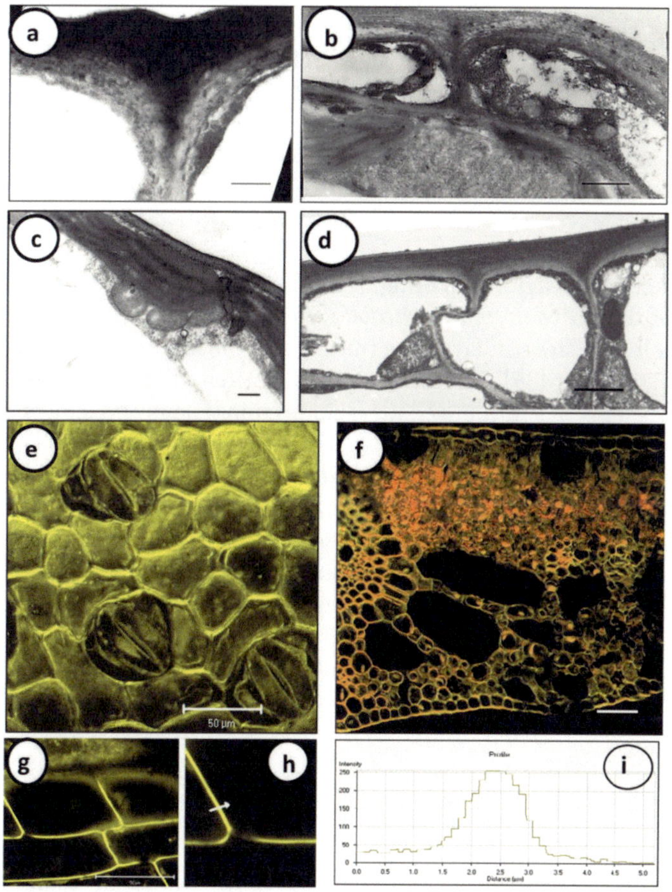

Figure 4.
Fragments of epidermal cells of surface above-water (a, c, d) and submerged (b) leaves of Sagittaria sagittifolia (a and b) and Trapa natans (c and d). In submerged leaves, the epidermal cell walls are thinner and loosened. Figures e, f, g, h, and i show the cytochemical fluorescence of lignin in cells of the above-water S. sagittifolia. Lignin has a yellow fluorescence: e, f, adaxial epidermis; g, h, i, abaxial epidermis. In figure i, the histogram of intensity lignin yellow is presented. Figure e shows 3D structure. Scale bars: a, b, c, d = 1 μm; e, f, g = 50 μm [26].

up to 2.5 μm in the area of the anticlinal walls, characterized by layering. The mesophyll cell walls are thin and have a typical pecto-cellulose structure. A similar thinning of the epidermal cell walls was found in the free-floating leaves of *T. natans* [53].

The mechanism of cell wall thinning and loosening, particularly in hydrophytes, has not been fully understood. There are several proposed models for cell wall loosening, including associations with decreased cell turgor, reduced activity of polysaccharide synthetases [54], apoplast acidification [55], and activation of phytohormones [56, 57]. It is revealed that phytohormones induce stretching and increase wall plasticity, depending on the apoplast pH [54, 56]. Additionally, phytohormones act as effectors of wall loosening. It has been established that one mechanism of cell wall loosening in higher plant stems is the participation of brassinosteroids, which regulate the expression of genes for loosening proteins, namely xyloglucan endotransglucosylases, and act at the transcriptional level [57].

Studies of lignin in the surface leaves of *S. sagittifolia* and *Trapa natans* showed that cells of the upper and lower epidermis contain lignin, with different relative content in the periclinal and anticlinal epidermal walls [53, 58]. Laser confocal microscopy revealed light yellow fluorescence of lignin in the epidermal and mesophyll cell walls, clearly visible in the leaf cells of the arrowhead leaves of *S. sagittifolia* (**Figure 4c–f**). The analysis showed that the relative lignin content in the walls of the upper epidermis was higher than in the walls of the lower epidermis. Additionally, the relative lignin content in the mesophyll cell walls was two to four times lower than in the walls of the upper epidermis. Biochemical analysis showed that the total lignin content in above-water leaves of *S. sagittifolia* was 26.5 ± 3.1 mg/g dry weight, whereas in the underwater leaves, its content was more than twice as high at 58.6 ± 4.9 mg/g dry weight.

The presence and distribution of lignin in the tissues of underwater and above-water leaves of the hydrophyte *S. sagittifolia*, as well as the increased lignin content in underwater leaves, are similar to the findings on the increase in lignin in flooded stems of *Ludviga repens* [39]. This phenomenon is attributed to the complex cycle of lignin synthesis, which is influenced by various endogenous and exogenous factors such as phytohormones, light, temperature, ambient gas concentration, and biotic stresses [37, 59]. Lignin provides mechanical support to plant tissues, enabling them to remain upright and withstand water pressure and wave action [31, 32]. Submerged plant organs must adapt to these environmental challenges, which affect their growth and structural parameters. Additionally, submerged leaves lack trichomes, stomata, and thick cuticles, making them more susceptible to pathogen invasion [25, 52]. In response to pathogen attacks, leaves and stems synthesize lignin as a defense mechanism [37]. Therefore, both exogenous and endogenous factors contribute to the increased lignin synthesis in the underwater leaves of the studied species of water nut and arrowroot.

A cytochemical analysis of *S. sagittifolia* revealed the presence of syringol and guajacyl in the cell walls of the upper and lower epidermis, palisade, and spongy parenchyma, as well as in the walls of the vessels of the conducting bundles (**Figure 5**) [45, 58].

The relative content of syringyl and guajacyl varied in the walls of the studied tissues. The highest values of guajacyl are found in the epidermis, palisade parenchyma, and vessels. The ratio of syringyl to guajacyl content (S/G) was highest in the abaxial epidermis, followed by the adaxial epidermis, spongy parenchyma, palisade parenchyma, and vessels. Notably, the periclinal walls had higher S/G values compared to the anticlinal walls, regardless of the type of epidermis. An increase in the S/G ratio strengthens the chemical barrier to protect the cell from water penetration and pathogen invasion [50]. Additionally, the S/G ratio indicates an increase in the mechanical strength of cell walls [51]. Leaves immersed in water are constantly in contact with surrounding aquatic microflora and numerous algae, and they can withstand constant water pressure and wave action [15]. Both the periclinal and anticlinal epidermal walls of underwater leaves protect the leaf cell surface from the exogenous effects of the aquatic environment by regulating the synthesis of lignin components.

3. Effect of drought on epidermis structure of plant leaves and monolignols content

Environmental stressors such as drought, salinity, salt fog, solar radiation, high light intensity, low nutrient availability, and high soil temperature affect the structure of a plant's leaves [60, 61]. These changes can impact the morphology of the leaves, which is directly regulated by primary stress signals that trigger mechanisms for adaptation, crucial for plant survival in stressful habitats [62]. Drought, for example,

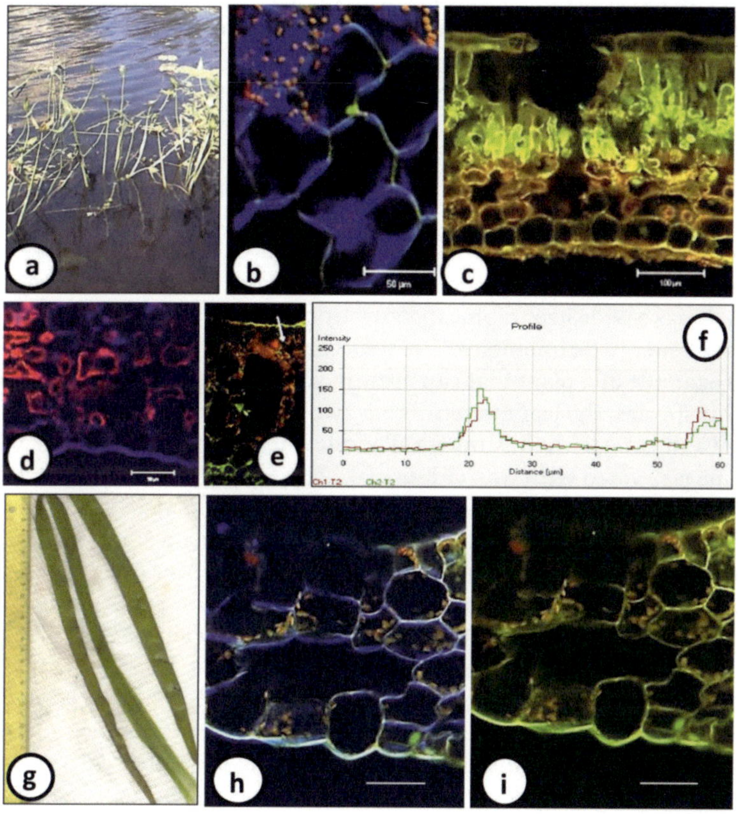

Figure 5.
The general view of the above-water (a) and submerged (g) leaves of Sagittaria sagittifolia. The cytochemical fluorescence of lignin components in the leaves is as follows: DPBA + syringyl complex fluoresces blue (b, d, h); DPBA + guajacyl complex fluoresces green (b, c, e, i); chlorophyll autofluorescence is red: b, h—syringyl + guajacyl; c, e, i—guajacyl; d—syringyl. Figure e, f shows the histogram of guajacyl fluorescence intensity (green line) and chlorophyll autofluorescence (red line). The fluorescence intensity is in relative units. The abscissa represents the distance (µm) scanned in figure e, which is shown by the white line. Scale bars: 100 µm (c) and 50 µm (b, d, h, i) [58].

can lead to a loss of leaf turgor, which is the plant's ability to maintain cell turgor pressure during dehydration and is a strong predictor of the plant's response to drought [63]. Various factors, including structural and functional changes in both the roots and leaves, help preserve leaf turgor, particularly in the leaf structure. Drought stress can reduce water potential and turgor, leading to inhibited photosynthesis, metabolism, and potential cell death [64]. Lignin plays a role in maintaining tissue water potential by regulating cuticular and transcriptional transpiration, thus reducing water consumption [65]. Vegetation in coastal sand dunes faces environmental stress such as drought, salinization, solar radiation, high light intensity, low nutrient availability, high temperatures, sand instability, and sea or river breezes [60, 61, 66]. These stressors affect the structure of plant organs [67].

Drought causes changes in leaf structure and these changes plays a significant role in leaf adaptation. Specific patterns in leaf morphology are regulated by primary stress signals, triggering mechanisms of structural and functional adaptation crucial for plant survival in stressful habitats [62]. Structural adaptations to drought conditions include water-storing parenchyma leaf blades, curling, deepening of stomata, and leaf pubescence. These adaptations help psammophytes maintain optimal water balance in plants. A study of 54 taxa in the Boraginaceae family found mineralized structures at the base of leaf trichomes in psammophytic species, as well as the presence of calcium and potassium ions [68].

For example, a study of species specific to sand dunes (Marina di Vecchiano coast, Italy), namely *Calystegia soldanella, Euphorbia paralias,* and *Otanthus maritimus*, showed that the leaves of *C. soldanella* and *E. paralias* had a flat shape, while those of *O. maritimus* were rolled up. The surface of the studied leaves of the three species was covered with a dense layer of glandular multicellular trichomes, which were located both on the entire surface and on the edges of the leaves [66, 69, 70]. *Phragmites australis* (Cav.) Trin. ex Steud is a plant species that can thrive in both water and drought conditions. It has a wide geographical distribution and is used in agriculture, construction, and energy technologies. The structure of the leaf epidermis of *P. australis* is a typical example of a plant that can grow in various environments. For terrestrial plants of this species, the leaf epidermis has two distinct zones: the stomata zone and the convex vault zone without stomata. The adaxial surface of the convex vault and stomata zone is covered with wax (**Figure 6**) [71]. The width of the vault zone ranges from 44.2 ± 3.1 to 65.0 ± 5.9 μm, while the width of the stomata zone ranges from 67 ± 4.1 to 100 ± 9.3 μm. Trichomes are present in the stomata zone, with a density of 376 ± 15 per mm2 area. The stomata are elongated-oval, with a density of 450 ± 39 on adaxial surfaces and 699 ± 53 per mm^2 area on abaxial surface, accordingly. The periclinal walls of guard cells are covered with wax granules. The abaxial epidermis also has a vault and a stomata zone. The widths of the stomata zone range from 50 ± 3.7 to 120 ± 10.1 μm, and the width of the convex vault zone is 53 ± 4.7 μm. Nipple-like short cells are present in the vault zone, covered with a thick layer of cuticle and wax structures. Guard cells and surrounding cells of stomata are encrusted with wax depositions. This investigation highlights the differences in leaf micromorphology and in the epidermis ultrastructure in the leaves of *P. australis* grown in different environments (in the water along the shore and on land, far from the water's edge) [71]. The author revealed that the density of stomata and trichomes in the leaf epidermis of terrestrial plants comparably was significantly lower than that in the air-water ecotype of this species. Similar changes in stomata density have also been found in rice leaves [72].

The study of both surfaces of the leaves of *Corynephorus canescens* grown in drought conditions showed the presence of non-glandular trichomes and wax. These

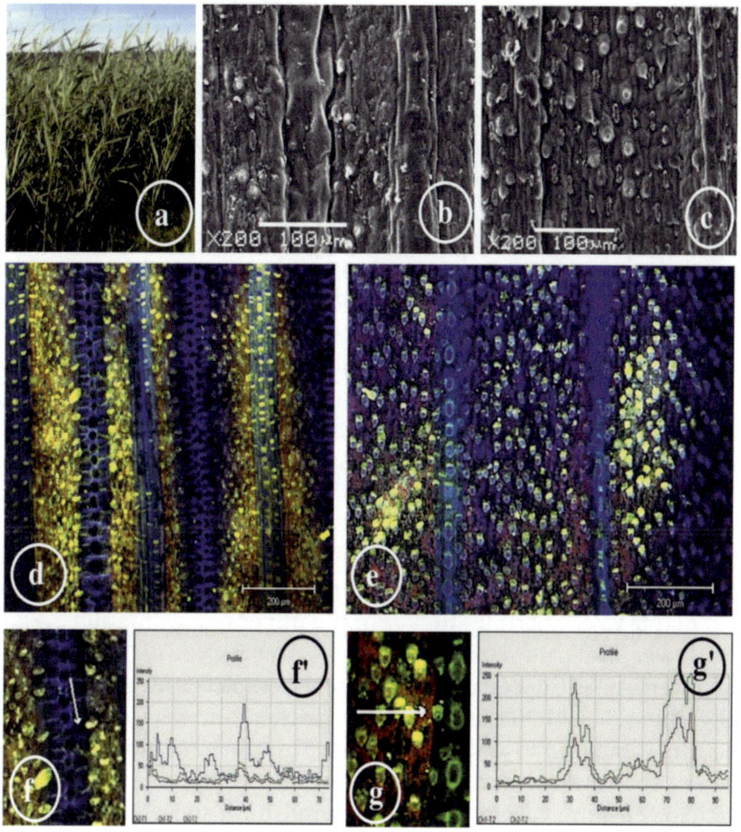

Figure 6.
The general view of Phragmites australis plants growing in terrestrial soil (a); the ultrastructure of the leaf adaxial (b) and abaxial surfaces (c). Micrographs of fluorescence of monolignols in leaf epidermis of P. australis are also presented, with syringyl exhibiting blue fluorescence and guajacyl exhibiting green fluorescence, while chlorophyll exhibits red autofluorescence. Monolignols are shown in figures d, e, f, and g, with syringyl and guajacyl in walls. Figures f and f' present a histogram of the fluorescence intensity of syringyl (blue line), guajacyl (green line), and chlorophyll autofluorescence intensity (red line), while figures g and g' present a histogram of the fluorescence intensity of guajacyl and chlorophyll. The ordinate represents fluorescence intensity in relative units, while the abscissa represents the distance (µm) scanned on figures f and g. This distance is shown as a white line on figures f and g. Figures d and f represent the adaxial epidermis, while figures g and g' represent the abaxial epidermis. Scale bars: 100 µm (figures b and c) and 200 µm (figures d and e) [71].

trichomes protect the aboveground organs of plants from attack by sucking insects and herbivores, as well as from mechanical damage [73]. Meanwhile, the epicuticular wax, which is found on the cuticle surface, plays a crucial role in structuring the surface at a sub-cellular scale. It self-assembles into crystals that act as the main barrier to water and small molecule loss from the cell, as well as reducing the uptake of liquids and molecules from the outside [74, 75]. The presence of wax and trichomes in the epidermal cells is an adaptive feature of heliophytes and psammophytes that grow in soil under drought conditions. Psammophytes, which grow on sand dunes, have developed mechanisms to resist and adapt to environmental variations [76–79]. Structural features of the epidermis of psammophytes include twisted leaf blades, changes in stomatal density, increased indumentum, thickened cuticles, and the synthesis of calcium inclusions in cells and intercellular spaces, which help maintain the optimal water balance of such plants during of droughts [80–82].

Trichome density also increases during drought [83]. A dense coating of trichomes changes the optical properties of the leaf surface and can reflect or absorb a certain wavelength of light. Trichomes, which capture a layer of air on the leaf surface, can also preserve heat and moisture [73]. In the cells of the epidermis and hypodermis of *Corynephorus canescens* leaves, we found inclusions of various shapes and sizes containing calcium and potassium. These inclusions can also be of different nature, as reported in the literature [69, 84].

The study of the structural and functional characteristics of *Phragmites australis* grown under drought conditions showed that such stress causes increased lignin synthesis [71].

A cytochemical and laser confocal microscopic study of monolignols in the leaf epidermis of the air-water and terrestrial ecotypes of *Ph. australis* revealed that a decrease in soil moisture during moderate drought led to an increase in monolignols content and an increase in the S/G ratio. We found that trichomes and the cells of the vault zone in the leaf epidermis are the main accumulators of monolignols. We hypothesize that the increase in monolignol content in the epidermal cells of the adaxial and abaxial leaves of terrestrial plants leads to a reduction in cuticular and stomatal transpiration, optimizing the water balance of plants grown in moderate drought (**Figure 6d–g**). It was established that the intensity of syringyl fluorescence in terrestrial plants was 1.22 times higher in the upper epidermis and 2.4 times higher in the lower epidermis compared to air-water *P. australis* plants. Similarly, the distribution intensity of fluorescence guajacyl was significantly higher in terrestrial plants, being nine times higher on the adaxial surface and 8.3 times higher on the lower epidermis compared to air-water plants. This phenomenon could be explained by the different environmental conditions experienced by terrestrial and air-water plants, which may lead to variations in the composition and distribution of lignin components.

The differences in the relative content of monolignols in different epidermal cells of *Phragmites australis* leaves can be explained by the involvement of peroxidase and laccase enzymes in lignin synthesis, which are key enzymes in the polymerization of lignin monomers [85]. Plant protection during drought occurs at the cell and molecular levels, including the increase of enzyme activity of CoA reductase and cinnamyl alcohol dehydrogenase, as well as the expression of relevant genes [86]. Based on these findings and our results, we can infer that the enhanced synthesis of monolignols (S and G) in the leaves of terrestrial *P. australis* plants is due to the activation of these enzymes.

It is known that the increase in lignin causes osmotic stress, leading to water loss or cell death and inhibiting plant growth and development [64]. Lignin can decrease water flow by cells, supporting osmotic balance and cell integrity [87–89]. Considering the literature and our results on the increased S/G ratio in the cells of conductive vessels and cells of the upper epidermis in leaves of the terrestrial ecotype of *P. australis*, we assume that the main factor influencing this indicator is relative drought and soil moisture. Previous experiments have shown that an increase in S/G quantity acts as a chemical barrier, intensifying cell defense from water penetration and pathogens [50] and increasing the mechanical durability of cells [51]. It is likely that the increased S/G value in the stomata and trichome of the stomata zone in aerial-aquatic and terrestrial reed plants increased the mechanical properties of the leaf epidermis of the studied species.

According to the literature, drought, similar to soil salinity, has various negative effects on plant growth, such as osmotic stress and inhibition of organ growth

[64, 87–89]. However, plants can adapt to unfavorable conditions by exhibiting phenotypic plasticity, which includes an increase in lignin synthesis and changes in leaf blade shape (twisting into tubes or folding) [90–92]. This helps the plant maintain optimal water balance [38, 89] and protects it from UV radiation. Lignin acts as an "aggressive" wall component that can displace other polysaccharides [93]. Researchers have proposed a biomechanical model demonstrating that crystalline cellulose is replaced by lignin in the stomatal end walls, serving to strengthen the wall.

Besides, lignin and its monomers are a chemical barrier that increases the protecting of cells from penetration of water and invasion by pathogens [50], and the sign (S/G) testifies to the increase of mechanical durability of cells [51]. The presence of lignin and crystalline cellulose in guard cells implies different biomechanical functions. Our experimental data suggests that in terrestrial plants, leaves of reeds receive a higher dose of ultraviolet radiation than leaves of reeds in plants growing in water. The high content of syringyl monolignol in terrestrial plants can serve as a marker for making strong bricks, while the increased content of other guajacyl monolignol in air-water plants can be a marker for getting soft paper from the reeds. The presence and ratio of monolignols in the cell walls of the epidermis and vessels correlate with early-received data for upland plants [94]. It is possible that the leaf cell walls of the plants studied also contain p-hydroxyphenyl, which plays a substantial role in lignin structure and can be determined using UV spectroscopy [95].

Thus, it was established that the changes at tissue and cellular levels are accompanied by an increase in lignin and cellulose in plants grown under drought stress. However, changes in lignin content in plants under the influence of the environment vary, with some experiments showing a decrease and others showing an increase. This may depend on the plant species and growth phase used for the study. Plant protection during drought occurs at both the cellular and molecular levels. Researchers have noted increased enzyme activity of CoA reductase and the expression of relevant genes in response to moderate water deficiency in the soil. This led to an increased synthesis of monolignols in the cell walls of *P. australis* leaf. The increase in syringyl and guajacyl content in the cell walls of epidermis cells of terrestrial plants may reduce cuticular and stomata transpiration, optimizing the water balance of plants grown in moderate drought and increasing plant resistance to drought.

4. The influence of cold on leaf lignin

The main causes of plant death from cold are direct or prolonged exposure to low temperatures and the impact on crop growth and development. Researchers categorize plants into three types based on their response to cold: sensitive to cooling, resistant to cold, and adapted to low temperatures [96]. Plants that overwinter in temperate climates can tolerate the freezing of a significant amount of water on the cell surface. Ice crystal formation begins when exposed to temperatures ranging from −1 to −3°C for one hour. The receptors for cold perception in a plant cell have not yet been identified, but it is assumed that the cold signal is perceived by converting it into a calcium signal, which triggers further processes [97].

A study of the structural and functional features of plants growing in the Arctic may provide an answer to the natural mechanisms of adaptation of the plant world to the effects of low temperatures. Polar plants are characterized by the inhibition of respiration in the light, with ATP production occurring during the day through photorespiration and dark respiration observed only at night. This has been shown in polar species such as *Betula nana* L. (dwarf birch) and *Salix pulchra* (tea vine) [98].

Cold shock proteins and genes, including desaturases, and dehydrins are hydrophilic and are secreted into the cell walls of epidermal tissues. They are then placed on the surface of ice crystals, inhibiting their further growth [99, 100]. Freezing water from the outside of the cell prevents the formation of ice crystals in the protoplast but leads to dehydration of the cell's macromolecular structures. This occurs due to increased outflow of water from the cell through the plasmalemma, resulting in changes to the composition of phospholipids. Water then goes into the intercellular spaces, causing severe dehydration of the protoplast [101, 102]. These events lead to a decrease in cell volume, deformation, and changes in the mechanical properties of cell walls [103–105].

Cold acclimatization leads to changes in the structure of the cell wall and the composition of polysaccharides, including thickening of the cell wall of the epidermis and alterations in the synthesis of cellulose, soluble polysaccharides, and pectins [105, 106]. Studies on oilseed rape seedlings exposed to cold conditions showed significant changes in the content of major polysaccharides [107]. The resistance of the cell walls of the epidermis to frost affects the protection of the protoplast from ice formation, resulting in changes in leaf structure and cell wall thickness in cold conditions [99, 108].

A study of the epidermis ultrastructure in the leaves of *Deschampsia antarctica* collected during the 20th Ukrainian Antarctic Expedition showed that the morpho-anatomical traits of the leaves collected on the islands Scua and Galindez are similar to those of the species from other regions of the Maritime Antarctic [109–112]. It was found that all the leaves of the studied species twisted in a tube-shaped curl. The epidermis of leaves of *D. antarctica* plants collected on the islands Scua and Galindez characterized the next ultrastructural signs. Oval cells of the adaxial epidermis were covered with an even, smooth cuticle layer (**Figure 7a** and **b**) [109]. It was shown that leaf blades were downy.

It was early established that low temperatures did not significantly affect the content of lignin, but there were changes in enzyme activity and an increase in the expression of genes coding for *C3H* protein and cytochrome *P450* monooxidase, which are involved in lignin synthesis [113]. An increase in lignin, cellulose, and hemicellulose content was also observed in poplar seedlings and monocots with a decrease in air temperature [114, 115]. The presence of two monolignins (syringyl and guajacyl) in *D. antarctica* leaf cell walls was revealed (**Figure 7c–e**) using cytochemical methods and confocal microscopy [109]. The study showed that the content of syringyl and guajacyl in the walls of the adaxial and abaxial epidermis was different, with the highest content of syringyl in the periclinal walls of the upper epidermis and the highest content of guajacyl in the inner walls of both epidermis, the walls of the mesophyll, and conducting bundles. The relative content of syringyl in epidermal cell walls exceeded the relative content of guajacyl by 6–8 times, indicating that the change in lignin components is essential for adaptive changes in response to cold conditions.

Summing up the numerous literature data and the results of our experimental studies, we conclude that regardless of the type of extreme environmental impact on plants, the cells of aboveground organs undergo increased synthesis of lignin.

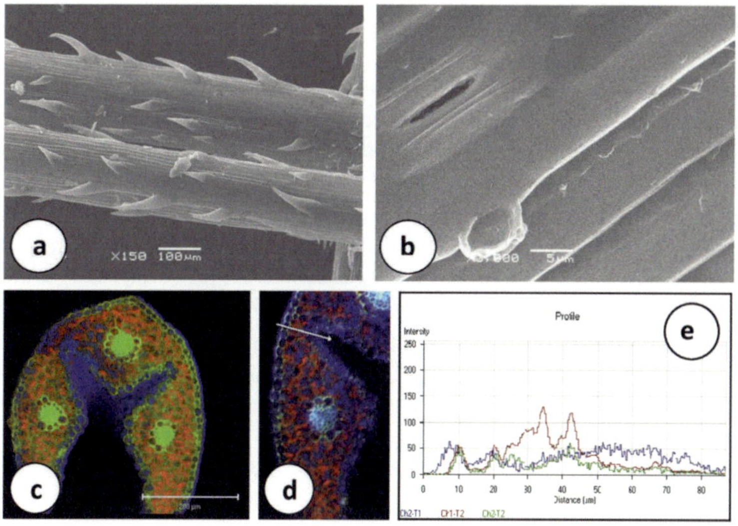

Figure 7.
Ultrastructure of epidermal cells of the leaves of Deschampsia antarctica (a and b) and cytochemical fluorescence of monolignols in D. antarctica leaves (c and d). The syringyl fluoresces in blue; the guajacyl fluoresces in green, and chlorophyll—in red. The histogram in figure e shows the fluorescence intensity of syringyl, guajacyl, and chlorophyll; the abscissa represents the distance (μm) scanned in figure d, which is shown by the white line. Scale bars is 100 μm (a) and = 5 μm (b) [109].

The content and composition of lignin depend on the type of stress and its duration. Cellulose is also synthesized in the cell walls in parallel, regardless of the type of stress. The crystalline form of cellulose, along with monolignins, creates a natural barrier to water transport by inhibiting apoplastic transport, as well as cuticular and stomatal transpiration [30, 115]. It is known that the amorphous zones of cellulose are involved in the transport and absorption of water molecules, while the crystalline form of cellulose does not have this ability [116]. We believe that it is extremely interesting and valuable for fundamental biology and applied sciences to establish the interaction of lignin oligomers with the crystalline form of cellulose-1β [117, 118]. This was experimentally demonstrated using the leaves of *Miscanthus giganteus* grass. The researchers have shown that the connection between the polysaccharide cellulose and lignin oligomers depends on the molecular weight of the oligomer; the higher the molecular weight of the monolignol, the stronger this connection. Recently, a group of scientists, including López-Albarrán et al. [118], used modeling to study the molecular interactions of oligolignols with cellulose I-β. They concluded that trilignol CA_βO4_SA_ββ_SA and dilignol CA_ ββ_CA were the lignin substructures that most contribute to cellulose adhesion.

In practice, this invention is used in the development of natural wood adhesives. In my opinion, such natural adhesives with lignin oligomers and cellulose can also be used in medicine, textile, and building materials industry. In our opinion, the leaves of reeds and cattails that grow in the coastal strips of European rivers, including the rivers of Ukraine, can serve as a source of such adhesives. Why reeds and cattails? Because the leaves of these plants, in addition to the monolignols (guajacyl and syringyl) and crystalline cellulose, also contain a lot of silicon, which easily absorbs and reflects sunlight. Such properties of natural materials are valuable for their further use in practice.

5. Conclusions

This review indicates that the leaf epidermis plays a role in the adaptation of plants to submergence and drought at structural and biochemical levels. The adaptation mechanisms of the epidermis in leaves to flooding are expressed in the absence of stomata and wax in the leaf epidermis, the loosening of cell walls, and the presence of cuticle pores. Meanwhile, the adaptation mechanisms of the leaf epidermis to drought are expressed in an increase of wax and cuticle, as well as an increase in trichome size and density. The questions involving the epidermal walls of submerged and drought plants are so varied and numerous that this field of plant morphology, anatomy, and cell biology will continue to attract the attention of many investigators to establish not only regularities of fundamental biology but also to use theoretical-experimental data for constructing agricultural plants with resistance to environmental stress.

The adaptation mechanisms of the epidermis of leaves to flooding and drought are also expressed in the synthesis of monolignols syringyl and guajacyl, the content of which depends on environmental conditions. Syringyl and guajacyl monolignols were detected in the leaves of air-water and terrestrial plants using cytochemical methods and laser confocal microscopy. The Pascal program for laser confocal microscopic study of monolignols in the leaf epidermis of hydrophytes revealed that the external environment of the leaf epidermis (water or air) can change the content of monolignols in the epidermis wall. It is established that the anticlinal walls of usual epidermal cells are the main accumulator of guajacyl monolignol. An increase in the content of guajacyl monolignol in the anticlinal walls of the epidermis leads to an increase in buoyancy and the optimization of plant life support on the water surface. It was found that the decrease in soil moisture leads to an increase in the content of monolignols (syringyl and guajacyl, S + G, and ratio of S/G) in the epidermis of plants under drought. An increase in syringyl and guajacyl content in the cell walls of epidermal cells of the leaves of terrestrial plants can reduce cuticular and stomata transpiration and lead to optimization of the water balance of plants grown in moderate drought.

Our experimental data show that the high content of syringyl monolignol, which gives high reed strength in terrestrial plants, can serve as a marker for the use of terrestrial reed plants for making strong bricks for commercial use. The increased content of other guajacyl monolignol in the leaves of air-water plants gives softness to the leaves and can be a marker for getting soft paper from the reeds grown on the banks of the river.

The tolerance mechanism of the leaf to cold is expressed through an increase in syringyl synthesis in the outer wall of the leaf epidermis. Given that syringol monolignol increases the mechanical stability of cells and acts as a barrier to water transport [51] and the action of ice crystals, we can assume that syringol in the periclinal walls of the epidermis of artificial grasses is an active component of plant cold resistance.

Acknowledgements

The author thanks Dr. Klimchuk D.O. for help during scanning electron microscopy of the samples.

Funding

The preparation and writing of the manuscript chapter (title: "Influence of Environmental Change on Monolignols and the Micromorphology of Leaf Epidermis in Hydrophytes and Terrestrial Plants" for Advances Lignin Technologies) was not funded by M.G. Kholodny Institute of Botany of the National Academy of Sciences of Ukraine.

Conflict of interest

The authors declare no conflict of interest.

Author details

Olena Nedukha
Department of Cell Biology and Anatomy, M.G. Kholodny Institute of Botany, National Academy of Sciences of Ukraine, Kyiv, Ukraine

*Address all correspondence to: o.nedukha@hotmail.com

References

[1] Dietz K-J, Hartung W. The leaf epidermis: Its ecophysiological significance. Progress in Botany. Fortschritte der Botanik. 1996;**57**:32-53. DOI: 10.1007/978-3-642-79844-3_3

[2] Mommer L, Visser EJW. Underwater photosynthesis in flooded terrestrial plants: A matter of leaf plasticity. Annals of Botany. 2005;**96**:581-589. DOI: 10.1093/aob/mci212

[3] LoPresti EF. Chemicals on plant surfaces as a heretofore unrecognized, but ecologically informative, class for investigations into plant defence. Biological Reviews. 2015;**91**:1102-1117. DOI: 10.1111/brv.12212

[4] Kerstein G. Cuticular water permeability and its physiological significance. Journal of Experimental of Botany. 1996;**47**:1813-1832. DOI: 10.1093/jxb/47.12.1813

[5] Werker E. Trichome diversity and development. Advances in Botanical Research. 2000;**31**:1-35. DOI: 10.1016/S0065-2296(00)31005-9

[6] Bickford CP. Eco-physiology of leaf trichomes. Functional Plant Biology. 2016;**43**:807-814. DOI: 10.1071/FP16095

[7] Dixon RR, Chen F, Guo D, Parvathi K. The biosynthesis of monolignols: A "metabolic grid", or independent pathways to guaiacyl and syringyl units? Phytochemistry. 2001;**57**:1069-1084. DOI: 10.1016/S0031-9422(01)00092-9

[8] Roland JC, Vian B. The wall of the growing cell: Its three dimension organization. International Review of Cytology. 1979;**61**:129-166. DOI: 10.1016/S0074-7696(08)61997-X

[9] Schuetz M, Benske A, Smith R, Watanabe Y, Tobimatsu Y, et al. Laccases direct lignification in the discrete secondary cell wall domains of protoxylem. Plant Physiology. 2014;**166**:798-807. DOI: 10.1104/pp.114.245597

[10] Vartapetian BB, Jackson MB. Plant adaptations to anaerobic stress. Annals of Botany. 1997;**79**:3-20. DOI: www.jstor.org/stable/42764824

[11] Doronin Y. Ocean Physics. St. Petersburg: Russian State Hydrometeorological University. Publ. RSHU; 2000. 340 p. (in Russ.)

[12] Madsen TV, Maberty SC. Diurnal variation in light and carbon limitation of photosynthesis by two species of submergered freshwater macrophytes with a differential ability to use bicarbonate. Freshwater Biology. 1991;**26**:175-187. DOI: 10.1111/j.1365-2427.1991.tb01727.x

[13] Sand-Jensen K, Pedersen MF, Nielsen SL. Photosynthetic use of inorganic carbon among primary and secondary water plants in streams. Freshwater Biology. 1992;**27**:283-293. DOI: 10.1111/J.1365-2427.1992.TB00540.X

[14] Mommer L, Pederson O, Visser EJW. Acclimation of a terrestrial plant to submergence facilitates gas exchange under water. Plant Cell and Environment. 2004;**27**:1281-1287. DOI: 10.1111/j.1365-3040.2004.01235.x

[15] Nedukha OM. Heterophylly in Plants. Kyiv: Publ.: Alt Press; 2011. 191 p. (in Ukrainian). ISBN 978-066-542-494-9

[16] Schönherr J. Resistance of plant surfaces to water loss: Transport

properties of cutin, suberin and associated lipids. In: Lange O., Nobel P., Osmond C., et al., editors. Physiological Plant Ecology. II. Encyclopedia of Plant Physiology. Heidelberg: Springer; 1982. 2B. p. 153-179.

[17] Schönherr J. Characterization of aqueous pores in plant cuticles and permeation of ionic solutes. Journal of Experimental Botany. 2006;**57**:2471-2491. DOI: 10.1093/jxb/erj217

[18] Bruni NC, Yang JP, Dengler NG. Leaf development plasticity of *Ranunculus aquatilis* in response to terrestrial and submerged environments. Canadian Journal of Botany. 1996;**74**:823-827. DOI: 10.1139/b96-103

[19] Rascio N, Cuccato F, Dalla Vecchia F, et al. Structural and functional features of the leaves of *Ranunculus trichphyllus* Chaix., a freshwater submerged macrophyte. Plant Cell and Environment. 1999;**22**:205-212. DOI: 10.1046/j.1365-3040.1999.00394.x

[20] Frost-Christensen H, Bolt Jorgensen L, Flotoo F. Species specificity of resistance to oxygen diffusion in thin cuticular membranes from amphibious plants. Plant, Cell and Environment. 2003;**26**:561-569. DOI: 10.1046/j.1365-3040.2003.00986.x

[21] Frost-Christensensen H, Floto F. Resistance to CO_2 diffusion in cuticular membranes of amphibious plants and the implication for CO_2 acquisition. Plant Cell and Environment. 2007;**30**:12-18. DOI: 10.1111/j.1365-3040.2006.01599.x

[22] Hoffmannbenning S, Kend H. On the role of abscisic-acid and gibberellin in the regulation of growth in rice. Plant Physiology. 1992;**99**:1156-1161. DOI: 10.1104/pp.99.3.1156

[23] Armstrong W, Brande R, Jackson MB. Mechanisms of flood tolerance in plants. Acta Botanica Neerlandica. 1994;**43**:307-358. DOI: 10.1111/j.1438-8677.1994.tb00756.x

[24] Jackson MB, Colmer TD. Response and adaptation by plants to flooding stress. Annals of Botany. 2005;**96**:501-505. DOI: 10.1093/aob/mci205

[25] Nedukha OM. Epidermis leaf structural responses of some aquatic plants to constant water environment. Advances of Agricultural Sciences. Problem Issue. 2010;**545**(545):169-178. DOI: 10.13140/RG.2.1.2174.0562

[26] Nedukha OM. Structural features of the cell walls in the leaves of *Sagittaria sagittifolia* and *Trapa natans*. Modern Phytomorphology. 2013;**3**:259-265. УДК 57.012.4:581.821

[27] Nedukha OM. Role of cell walls in plant adaptation to flooding. Journal of Biology and Nature. 2017;**7**:38-62. ISSN: 2395-5376 (print), 2395-5384

[28] Yates D, Hutley LB. Polar uptake of water by wet leaves of *Sloanea woollsii* an Australian subtropical rainforest trees. Australian Journal of Botany. 1995;**43**:157-167. DOI: 10.1071/BT9950157

[29] Leisola M, Pastinen O, Axe DD. Lignin—Designed. Bio-complexity.org. 2012;**3**:1-11. DOI: 10.5048/BIO-C.2012.3.c

[30] Panda D, Sarkar RK. Structural carbohydrates and lignifications associated with submergence tolerance in race (*Oryza sativa* L.). Journal of Stress Physiology and Biochemistry. 2013;**9**:299-306. ISSN: 1997-0838

[31] Tyree MT, Cheung YNS. Resistance to water flow in Fagus grandifolia leaves. Canadian Journal of Botany. 1997;**55**:2591-2599. DOI: 10.1139/b77-296

[32] Niklas KJ. Plant Biomechanics: An Engineering Approach to Plant Form and Function. University of Chicago Press; 1992. 622 p

[33] Wilson K, White DJB. Reaction wood: Its structure, properties and functions. In: Wilson K, Stobart WDJB, editors. The Anatomy of Wood: Its Diversity and Variability. London, UK; Stobart; 1986. pp. 222-250. ISBN 10: 0854420339/ISBN 13: 9780854420339

[34] Joseleau J-P, Imai T, Kuroda K, Ruel K. Detection in situ and characterization of lignin in the G-layer of tension wood fibres of *Populus deltoides*. Planta. 2004;**219**:338-345. DOI: 10.1007/s00425-004-1226-5

[35] Nedukha OM. The participation of wall monolignols in leaf tolerance to nature flooding of hydrophytes. Advances in Bioscience and Biotechnology. 2015;**6**:546-555. DOI: 10.4236/abb.2015.68058

[36] Nedukha OM. Cytochemical study of monolignols in cell walls of hydrophytes leaves. Ukrainian Journal of Botany. 2015;**72**:385-392. (In Ukrainian). DOI: nbuv.gov.ua/UJRN/UBJ_

[37] Moura JC, Bonine CA, Viana J, Dornelas MC, Mazzafera P. Abiotic and biotic stresses and changes in the lignin content and composition in plants. Journal Integrative Plant Biology. 2010;**52**:360-376. DOI: 10.1111/j.1744-7909.2010.00892.x

[38] Hilal M, Parrado M, Rosa M, Gallardo M, Orce L, et al. Epidermal lignin deposition in *quinoa* cotyledons in response to UV-B radiation. Photochemistry and Photobiology. 2004;**79**:205-210. DOI: 10.1562/0031-8655(2004)079 <0205:ELDIQC>2.0.CO;2

[39] Little ST. Adaptation and acclimation of populations of *Ludwigia repens* to growth in high- and lower CO_2 springs [a dissertation presented to the graduate school of the University of Florida in partial fulfillment of the requirements for the degree of doctor of philosophy]. USA: Univ. Florida; 2003. pp. 1-157

[40] Vanholme R, De Meester B, Ralph J, Wout BW. Lignin biosynthesis and its integration into metabolism. Current Opinion in Biotechnology. 2019;**56**:230-239. DOI: 10.1016/j.copbio.2019.02.018

[41] Alain-M B. Towards an understanding of the supramolecular organization of the lignified wall. In: Rose JKC, editor. The Plant Cell Wall. Ithaca, NY, USA: Dep. Plant Biol. Cornell Univ.; 2001. pp. 155-182

[42] Lai YZ, Funaoka M, Chen HT. Chemical heterogeneity in woody lignins. In: Argyropoulos DS, editor. Advances in Lignocellulosics Characterization. Atlanta: TAPPI Press; 1999. pp. 43-53

[43] Chabannes M, Ruel K, Yoshinaga A. *In situ* analysis of lignins from tobacco transgenic lines down-regulated for several enzymes of monolignol synthesis reveals a differential impact of transformation on the spatial patterns of lignin deposition at the cellular and subcellular levels. Plant Journal. 2001;**28**:271-282. DOI: 10.1046/j.1365-313x.2001.01159.x

[44] Ancibor E. Systematic anatomy of vegetative organs of the Hydrocharitaceae. Botanical Journal of the Linnean Society. 1970;**78**:237-266. DOI: 10.1111/j.1095-8339.1979.tb02570.x

[45] Nedukha OM. Heterophylly in *Sagittaria sagittiflia*. II. Surface ultrastructure of above-water and submerged leaves. Ukrainian Botanical Journal. 2012;**69**:596-603. (In Ukrainian)

[46] Bercu R. Some histological and cytological features on four hydrophytes

occurring in the Danube Delta (Romania). Contributii Botanice, (Grădina Botanică "Alexandru Borza" Cluj-Napoca). 2003;**38**:127-134

[47] Sanduescu EB, Stavrescu-Bedivan MM, Scaeteanu GV, Schiopu T. Morpho-anatomic features and chemical compounds in some aquatic plant species – preliminary data. Scientific Papers. Series A. Agronomy. 2014;**LVII**:441-447. ISSN 2285-5785, ISSN CD-ROM 2285-5793

[48] Jung J, Lee JSC, Chu H-K. Anatomical patterns of aerenchyma in aquatic and wetland plants. Journal of Plant Biology. 2008;**51**:428-439. DOI: 10.1007/bf03036065

[49] Vogel S. Contributions to the functional anatomy and biology of *Nelumbo nucifera* (Nelumbonaceae) I. Pathways of air circulation. Plant Systematics and Evolution. 2004;**249**:9-25. DOI: 10.1007/s00606-004-0201-8

[50] Menden B, Kohlhoff M, Moerschbacher BM. Wheat cell accumulate a syringil-rich lignin during the hypersensitive resistance response. Phytochemistry. 2007;**68**:513-529. DOI: 10.1016/j.phytochem.2006.11.011

[51] Christiernin M. Composition of lignin in outer cell-wall layers [PhD thesis]. Stockholm: Division of Wood Chemistry and Pulp Technology, Royal Institute of Technology, TRITA-FPT-Report; 2006. p. 153. ISSN1652-2443 ISRN KTH/FPT/R-2006/16-SE

[52] Nedukha OM. Cell Wall and Environmental Factors. Kyiv: Alt Press; 2015. 288 p. (In Ukrainian. ISBN 978-966-542-575-5)

[53] Nedukha OM, Kordyum EL. The plasticity of anatomical structure and cell wall lignin in *Trapa natans* adaptation to nature flooding. Annals of the Romanian Society for Cell Biology. 2016;**21**:27-34. DOI: annalsofrscb.ro/index.php/journal/article/view/13

[54] Cosgrove DJ. Molecular regulation of plant cell wall extensibility. Gravitational and Space Biology Bulletin. 1998;**11**:61-70

[55] Rayle DL, Cleland RE. The acid growth theory of auxin-induced cell elongation. Plant Physiology. 1992;**99**:1271-1274. DOI: 10.1104/pp.99.4.1271

[56] Cosgrove DJ. Wall structure and wall loosening. A look backwards and forwards. Plant Physiology. 2001;**125**:131-134. DOI: 10.1104/pp.125.1.131

[57] Clouse SD. Molecular genetic analysis of brassinosteroid action. Physiology Plantarum. 1997;**100**:702-709. DOI: /10.1111/j.1399-3054.1997.tb03077.x

[58] Nedukha OM. The participation of cell wall polysaccharide in cellular mechanisms of leaf tolerance to nature flooding of plants In: M.T. Grazeslak. Plant Functioning Under Environmental Stress. Cracow, Poland: DRUKROL LTD; 2013. p. 137-152. ISBN 978.83.86978-30-7

[59] Rastogi S, Dwivedi U. Manipulation of lignin in plants with special reference to O-methyltransferase. Plant Science. 2008;**174**:264-277. DOI: 10.1016/j.plantsci.2007.11.014

[60] Maun MA. Adaptations of plants to burial in coastal sand dunes. Canadian Journal of Botany. 1997;**6**:713-738. DOI: 10.1139/b98-058

[61] Grigore MN, Toma C. Histo-anatomical strategies of *Chenopodiaceae* halophytes; adaptive, ecological and evolutionary implications. WSEAS Transactions on Biology

and Biomedicine. 2007;**12**:204-218. ISSN:1109-9518

[62] Catoni R, Gratani L. Morphological and physiological adaptive traits of Mediterranean narrow endemic plants: The case of *Centaurea gymnocarpa* (Capraia Island, Italy). Flora. 2013;**208**:174-183. DOI: 10.1016/j.flora.2013.02.010

[63] Zhu SD, Chen YJ, Ye Q, He PC, Liu H, et al. Leaf turgor loss point is correlated with drought tolerance and leaf carbon economics traits. Tree Physiology. 2018;**38**:658-663. DOI: 10.1093/treephys/tpy013

[64] Chaves MM, Flexas J, Pinheiro C. Photosynthesis under drought and salt stress: Regulation mechanisms from whole plant to cell. Annals of Botany. 2009;**103**:551-560. DOI: 10.1093/aob/mcn125

[65] Hu Y, Li WC, Xu YQ, Li GJ, Liao Y, et al. Differential expression of candidate genes for lignin biosynthesis under drought stress in maize leaves. Journal of Applied Genetics. 2009;**50**:213-223. DOI: 10.1007/BF03195675

[66] Hesp P. Ecological processes and plant adaptations on coastal dunes author links open overlay panel. Journal of Arid Environments. 1991;**21**:165-191. DOI: 10.1016/S0140-1963(18)30681-5

[67] Le Bagousse-Pinguet Y, Forey E, Touzard B, Michalet R. Disentangling the effects of water and nutrients for studying the outcome of plant interactions in sand dune ecosystems. Journal of Vegetation Science. 2013;**24**:375-383. DOI: 10.1111/j.1654-1103.2012.01462.x

[68] Selvi F, Bigazzi MM. Leaf surface and anatomy in Boraginaceae tribe Boragineae with respect to ecology and taxonomy. Flora. 2001;**196**:269-285. DOI: 10.1016/S0367-2530(17)30056-7

[69] Hua Cong Ci, Xing Dong He, Rong Li, Wei Wu, PingPing Xue, et al. Characteristics of plant calcium fractions for 25 species in Tengger Desert. Sciences in Cold and Arid Regions. 2010;2:168-174. DOI: www.researchgate.net/publication/267692985

[70] Perronea R, Salmeri C, Brullo S, Colombo P. What do leaf anatomy and micro-morphology tell us about the psammophilous *Pancratium maritimum* L. (Amaryllidaceae) in response to sand dune conditions? Flora. 2015;**213**:20-31. DOI: 10.1016/j.flora.2015.03.001

[71] Nedukha OM. Micromorphology and monolignols of leaf epidermis in *Phragmites australis* (Poaceae) of air-aquatic and terrestrial ecotypes. Protoplasma. 2021;**258**:949-961. DOI: 10.1007/s00709-021-01614-4

[72] Caine R, Yin X, Sloan J, Harrison E, Umar Mohammed U, et al. Rice with reduced stomatal density conserves water and has improved drought tolerance under future climate conditions. New Phytologist. 2019;**221**:371-384. DOI: 10.1111/nph.15344

[73] Peter AJ, Shanower TO, Romeis J. The role of plant trichomes in insect resistance: A selective review (International Crops Research Institute for the semi-arid tropics). Journal: Phytophaga. 1995;7:41-63. Available from: https://www.scirp.org›refer

[74] Hauke V, Schreiber L. Ontogenetic and seasonal development of wax composition and cuticular transpiration of ivy (*Hedera helix* L.) sun and shade leaves. Planta. 1998;**207**:67-75. DOI: 10.1007/s004250050456

[75] Barlott W, Mattias M, Bharat B, Koch K. Plant surface: Structures and functions for biomimetic innovations. Nano-Micro Letters. 2017;**9**:1-40. DOI: 10.1007/s40820-016-0125-1

[76] Ashraf M, Harris PJC. Photosynthesis under stressful environments: An overview. Photosynthetica. 2013;**51**:163-190. DOI: 10.1007/s11099-013-0021-6

[77] Ruocco M, Bertoni D, Sarti G, Ciccarelli D. Mediterranean coastal dune systems: Which abiotic factors have the most influence on plant communities? Estuarine, Coastal and Shelf Science. 2014;**149**:213-222. DOI: 10.1016/j.ecss.2014.08.019

[78] Futorna OA, Badanina VA, Zigalova CL. Ecologo-anatomic characterization of some species *Tragopogon* (Asteraceae) of Ukraine Flora. Biosystems Diversity. 2017;**25**:274-281. (In Ukrainian). DOI: 10.15421/011742

[79] Nedukha OM. Appearance of phenotypic plasticity of leaves in psammophyte Corynephorus canescens during flooding. Plant Introduction. 2021;**91**(92):24-35. DOI: 10.46341/PI2021011

[80] Abd E, Mona E, Abo-Alatta OK, Habib SA, Abd Elba Ola H. The anatomical features of the desert halophytes *Zygophyllum album* L.F. and *Nitraria retusa*. Annals of Agricultural Science. 2016;**61**:1-8. DOI: 10.1016/j.aoas.2015.12.001

[81] Hameed BH, Krishni RR, Sata SA. A novel agricultural waste adsorbent for the removal of cationic dye from aqueous solutions. Journal of Hazardous Materials. 2009;**162**:305-311. DOI: 10.1016/j.jhazmat.2008.05.036

[82] Jianu LD, Bercu R,Popoviciu DR. *Silene thymifolia* Sibth. et Sm. (Caryophyllaceae) – A vulnerable species in Romania: Anatomical aspects of vegetative organs. Notulae Scientia Biologicae. 2021;**13**:1-7. DOI: 10.15835/nsb13110875

[83] Bourland FM, Hornbeck JM, McFall AB, Calhoun SD. A rating system for leaf pubescence of cotton. The Journal of Cotton Science. 2003;**7**:8-15. Available from: http://journalofcotton.org

[84] Ciccarelli D, Forino L, Costantina M, Balestri M, Maria PA. Leaf anatomical adaptations of *Calystegia soldanella*, *Euphorbia paralias* and *Otanthus maritimus* to the ecological conditions of coastal sand dune systems. Caryologia. 2009;**62**:142-151. DOI: 10.1080/00087114.2004.10589679

[85] Berthet S, Demontcaulet N, Pollet B, Bidzinski P, Cezard L, et al. Disruption of *LACCASE4* and 17 results in tissue-specific alterations to lignification of *Arabidopsis thaliana* stems. The Plant Cell. 2011;**23**:1124-1137. DOI: ff10.1105/tpc.110.082792ff. ffhal-01000145f

[86] Fan L, Linker R, Gepstein S, Tanimoto E, Yamamoto R, et al. Progressive inhibition by water deficit of cell wall extensibility and growth along the elongation zone of maize roots is related to increased lignin metabolism and progressive stelar accumulation of wall phenolics. Plant Physiology. 2006;**140**:603-612. DOI: DOI; 10.1104/pp.105.073130

[87] Agrawal A. Induced responses to herbivory and increased plant performance. Science. 1998;**279**:1201-1202. DOI: 10.1126/science.279.5354.1201

[88] Begovic L, Abičić I, Alojzije L, Lepeduš H, Cesar V, et al.

Lignin synthesis and accumulation in barley cultivars differing in their resistance to lodging. Plant Physiology and Biochemistry. 2018;**133**:142-148. DOI: 10.1016/j.plaphy.2018.10.036

[89] Monties B, Fukushima K. Occurrence, function and biosynthesis of lignins. In: Hofrichter M, Steinbuchel A, editors. Biopolymers. Lignin, Humic Substances and Coal. Vol. 1. Weinheim: Wiley; 2001. pp. 1-64. DOI: 10.1002/3527600035.bpol1001

[90] Micol JL, Hake S. The development of plant leaves. Plant Physiology. 2003;**131**:389-394. DOI: 10.1104/pp.015347

[91] Kadioglu A, Terzi R. A dehydration avoidance mechanism: Leaf rolling. The Botanical Review. 2007;**73**:290-302. DOI: 10.1663/0006-8101(2007)73[290:ADAMLR]2.0.CO;2

[92] Candela H, Johnston R, Gerhold A, Foster T, Hake S. The milkweed pod 1 gen encode a KANADI protein that is required for abaxial/adaxial patterning in maize leaves. The Plant Cell. 2008;**20**:2073-2087. DOI: 10.1105/tpc.108.059709

[93] Shtein H, Shelef Y, Marom Z, Zelinger E, Schwartz A, et al. Stomatal cell wall composition: Distinctive structural patterns associated with different phylogenetic groups. Annals of Botany. 2017;**119**:1021-1033. DOI: 10.1093/aob/mcw275

[94] Baucher M, Monties B, Van Montagu M, Boerjan W. Biosynthesis and genetic engineering of lignin. Critical Reviews in Plant Sciences. 1998;**17**:125-197. DOI: 10.1080/ 07352689891304203

[95] Fath L, Arno Frьhwald A, Koch G. Distribution of lignin in vascular bundles of coconut wood (*Cocos nucifera*) by cellular UV-spectroscopy and relationship between lignification and tensile strength in single vascular bundles. Holzforschung. 2014;**68**:915-925. DOI: www.researchgate.net/publication /269989221

[96] Thakur P, Nayyar H. Facing the cold stress by plants in the changing environment: Sensing, signaling, and defending mechanisms. In: Tuteja N, Gill SS, editors. Plant Acclimation to Environmental Stress. New York: Springer Science, Business Media; 2013. pp. 29-69

[97] Miura K, Furumoto T. Cold signaling and cold response in plants. International Journal of Molecular Sciences. 2013;**14**:5312-5337. DOI: 10.3390/ijms14035312

[98] Heskel M. Environmental Controls of Foliar Respiration in Arctic Tundra Plants. Submitted in Partial Fulfillment of the Requirements for the Degree of Doctor of Philosophy in the Graduate School of Arts and Sciences. Columbia University; 2013. 222 p

[99] Griffith M, Brown G. Cell wall deposition in winter rye *Secale cereale* L. "puma" during cold acclimatization. Botanical Gazette. 1982;**143**:486-490. Available from: www.jstor.org/stable/2474764

[100] Griffith M, Lumb C, Wiseman SB. Antifreeze proteins modify the freezing process in planta. Plant Physiology. 2005;**138**:330-340. DOI: 10.1104/pp.104.058628

[101] Steponkus PL, Uemura M, Webb MS. A contrast of the cryostability of the plasma membrane of winter rye and spring oat-two species that widely differ in their freezing tolerance and plasma membrane lipid composition. In:

Steponkus PL, editor. Advances in Low-Temperature Biology. Vol. 2. London: JAI Press; 1993. pp. 211-312

[102] Steponkus PL, Uemura M, Webb MS. Membrane destabilization during freeze-induced dehydration. Current Topics Plant Physiology. 1993;**10**:37-47. DOI: escholarship.org

[103]] Pearce RS. Extracellular ice and cell shape in frost-stressed cereal leaves: A low-temperature scanning-electron-microscopy study. Planta. 1988;**175**: 313-324. DOI: 10.1007/BF00396336

[104] Pearce RS, Ashworth EN. Cell shape and localization of ice in leaves of overwintering wheat during frost stress in the field. Planta. 1992;**188**:324-331. DOI: 10.1007/BF00192798

[105] Pearce Rogen S. Plant freezing and damage. Annals of Botany. 2001;**87**: 417-424. DOI: 10.1006/anbo.2000.1352

[106] Wallner SJ, Wu M, Anderson-Krengel SJ. Changes in extracellular polysaccharides during cold acclimation on cultured pear cells. Journal of the American Society for Horticultural. 1986;**111**:769-773

[107] Kubacka-Zebalska M, Kacperska A. Low temperature-induced modifications of cell wall content and polysaccharide composition in leaves of winter oilseed rape (*Brassica napus* L. var. *oleifera* L.). Plant Science. 1999;**148**:59-67

[108] Rajashekar C, Lafta A. Cell-wall changes and cell tension in response to cold acclimation and exogenous abscisic acid in leaves and cell cultures. Plant Physiology. 1996;**111**:605-612. DOI: 10.1104/pp.111.2.605

[109] Kordyum EL, Nedukha OM, Ovcharenko YV, Jadko SI, Ivanenko GF, et al. Anatomical and functional features of *Deschampsia antarctica* (Poaceae) leaf blade growing on the Argentine islands. Ukrainian Antarctic Journal. 2017;**16**:143-149. DOI: 10.33275/1727-7485.16.2017.72

[110] Gielwanowska I, Szczuka E, Bednara J, Gorecki R. Anatomical features and ultrastructure of *Deschampsia antarctica* (Poaceae) leaves from different growing habitats. Annals of Botany. 2005;**96**:1109-1119. DOI: 10.1093/aob/mci262

[111] Romero M, Casanova A, Iturra G, Reyes A, Montenegro G, et al. Leaf anatomy of *Deschampsia antarctica (Poaceae)* from the maritime Antarctic and its plastic response to changes in the growth conditions. Revista Chilena de Historia Natural. 1999;**72**:411-425

[112] Szczuka E, Gielwanowska I, Leszczuk A, Domaciuk M, Pietrusiewicz J, et al. Specific ultrastructure of the leaf mesophyll cells of *Deschampsia antarctica* Desv. (Poaceae). Annales Universitatis Mariae-Curie-Sklodowska, Sectio C – Biologia. 2013;**8**:25-33. DOI: 10.2478/v10067-012-0031-y

[113] El Kayal W, Keller G, Debayles C. Regulation of tocopherol biosynthesis through transcriptional control of tocopherol cyclase during cold hardening 873 in *Eucalyptus gunnii*. Physiology Plantarum. 2006;**126**:212-223. DOI: 10.3389/fpls.2021.743993

[114] Ford CW, Morrison IM, Wilson JR. Temperature effects on lignin, hemicellulose and cellulose in tropical and temperate grasses. Australian Journal of Agricultural Research. 1979;**30**:621-633. DOI: 10.1071/AR9790621

[115] Hausman JF, Evers D, Thiellement H, Jouve L. Compared responses of poplar cuttings and *in vitro* raised shoots to short-term chilling treatments. Plant Cell Reports.

2000;**19**:954-960. DOI: 10.1007/s002990000229

[116] Brown RM. The biosynthesis of cellulose. Journal of Macromolecular Science. 1996;**33**:1345-1373. DOI: 10.1080/10601329608014912

[117] Sedano-Mendoza M, Lopez-Albarran P, Pizzi A. Natural lignans as adhesives for cellulose: Computational interaction energy *vs* experimental results. Journal of Adhesion Science and Technology. 2010;**24**:1769-1786

[118] Lopez-Albarran P, Pizzi P, Navarro-Santos P, Hernandez-Esparza R. Oligolignols within lignin-adhesive formulations drive their Young's modulus: A ReaxFF-MD study. International Journal of Adhesion and Adhesives. 2017;**78**:227-233. DOI: 10.1016/j.ijadhadh.2017.08.003

Chapter 2

Deep Eutectic Solvents for Lignin Isolation

Tsegaye Abera Mekuriaw and Belachew Zegale Tizazu

Abstract

Deep eutectic solvents (DESs) are recently developed green solvents displaying characteristics, including simple production, reliability, chemical stability, low volatility, and biocompatibility that has caught the eye of researchers recently. This book chapter makes an in-depth review of the recent study on deep eutectic solvents, including its classifications, preparation methods, physicochemical characteristics, factors affecting DES pretreatment, and their applications. Based on the type of DES, molar ratio, type of biomass, reaction time, and reaction temperature, DESs can modify lignin characteristics and satisfy certain criteria. In addition, in this review, an investigation of the function of DESs in lignocellulose pretreatment and lignin isolation was also included. DESs dissolve lignin and provide extracted lignin with characteristics that make it suitable for further processing. This book chapter also discusses recent improvements in lignin extraction via biomass fractionation employing DESs and discusses DES-lignin valorization routes, depolymerization and functionalization.

Keywords: deep eutectic solvent, preparation techniques, physicochemical characteristics, lignin isolation, valorizations of DES lignin

1. Introduction

Lignin is an abundant lignocellulosic biomass next to cellulose, which makes up about 30% of organic carbon on earth [1]. Despite the availability and properties, the recalcitrant nature of the lignin structure makes it the most underutilized lignocellulose [2]. Lignin can be used for the production of various chemicals, bioenergy, biomaterials, polymers, and additives, to name a few [3]. However, a significant amount of lignin is left as waste or burned to generate heat [4]. Thus, from the biorefinery concept with a view to enhance the profitability, valorization of lignin with an efficient fractionation technique is the key to maximize biomass valorization and minimize waste [5].

The isolation of lignin from lignocellulosic biomass can be performed using solvents like dilute acid, alkali solvent, and organic solvents. However, these solvents are not suitable for the fractionation of lignin since they produce condensed lignin, despite the costs and environmental impacts [6]. Recently, researchers have investigated DES as the most efficient solvent for lignin isolation.

Deep eutectic solvents are green solvents prepared from binary or ternary combinations of different organic or inorganic chemicals, which are Lewis or Bronsted acids and bases, with anionic and cationic species through hydrogen bonding [7, 8], as shown in **Figure 1**. Because of its distinct and tailorable physicochemical qualities, such as low volatility, high solubility, biodegradability, cheap cost, environmental friendliness, and structural feasibility, deep eutectic solvent (DES) is considered as a new generation of green solvents and so on, and promoted their widespread usage in the effective delignification of various biomass feedstocks [9], as indicated in **Figure 2**.

Deep eutectic solvents are generally classified into four types: type (I, II, III), and type (IV), depending on the source of the complexing agent used.

The combinations of deep eutectic solvent type can be represented by the general equation $Cat^{+}X^{-}zY$, where Cat^{+} can be an ammonium, phosphonium, or sulfonium, X can be a Lewis base, mostly a halide anion, whereas Y can be a Lewis or Bronsted acid, and z denotes the number of molecules of Y [10]. Thus, in general, Type I, II, and III DESs can be prepared by mixing quaternary ammonium salt with metal chloride, metal chloride hydrate, and hydrogen bond donors, respectively. Whereas type IV is prepared by mixing metal chloride with hydrogen bond donor (HBD) [11].

The preparation of DES can be carried out by heating, grinding, evaporation, freeze-drying, microwave, and ultrasound-assisted methods. Heating and grinding

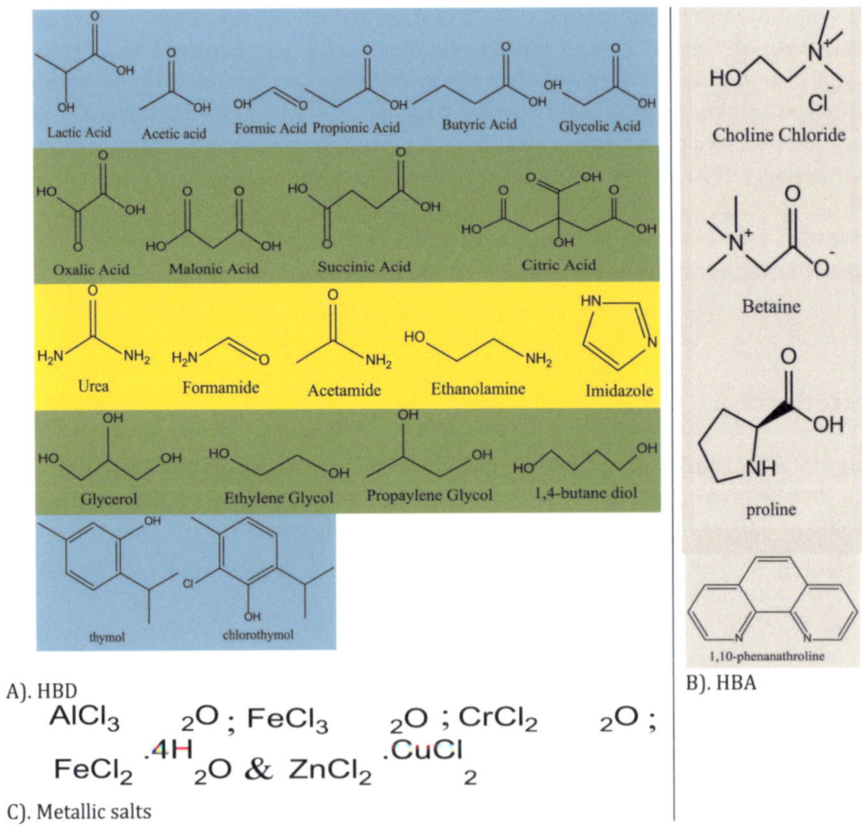

Figure 1.
Some commonly used organic or inorganic chemicals as hydrogen bond donors (HBDs), hydrogen bond acceptors (HBAs), and as metallic salts for DES synthesis.

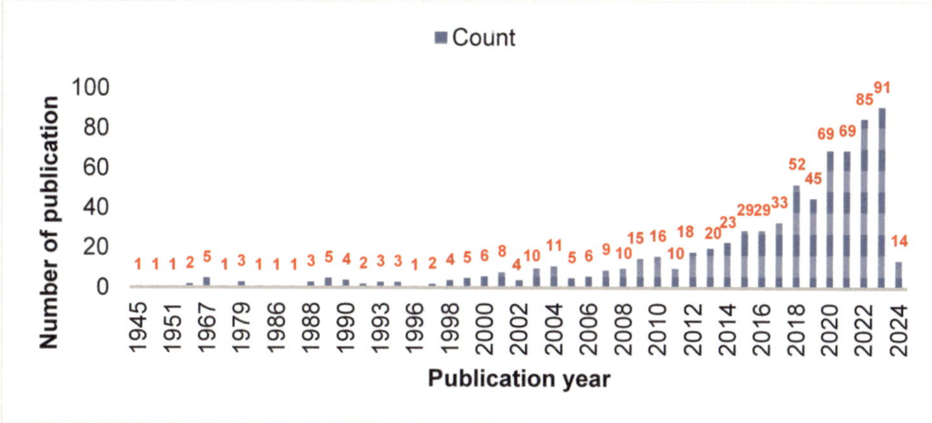

Figure 2.
Researches in the areas of deep eutectic solvent for lignin isolation.

methods are more commonly used than recently developed methods to overcome time and energy consumption, such as microwave-assisted and ultrasonic-assisted synthesis methods [11, 12]. However, ultrasound-assisted preparation of DES is the simplest and fastest preparation method, which is also chosen for the isolation of lignin from natural lignocellulosic materials in short time [13].

Earlier, processing of lignocellulose lignin was seen as complex and difficult due to lack of effective lignin extraction techniques. Thus, the valorization of lignocellulosic biomass typically, lignin became a problem. To overcome these problems, researchers found DESs are efficient and green methods for the extraction of lignocellulosic biomass.

The aim of this book chapter is to assess the most recent works on deep eutectic solvent, including its description, properties, synthesis and characterization, its role in lignocellulosic biomass isolation, factors affecting the isolation of lignin using DES, and the valorization routes of DES lignin.

2. Deep eutectic solvents

2.1 Deep eutectic solvent classification

Deep eutectic solvents (DESs) can be categorized, based on their nature of complexing agents as (type I, II, III, IV, and type V), pH (acidic, basic, and neutral), and miscibility in water (hydrophobic and hydrophilic) [14], which help to differentiate one DES from others. The majority of DESs are categorized under the nature of the complexing agent, which can be represented by the general formula $Cat^{+}X^{-}zY$, where Cat^{+} can be an ammonium, phosphonium, or sulfonium, X can be a Lewis base, mostly a halide anion, whereas Y can be a Lewis or Bronsted acid, and z denotes the number of molecules of Y [7, 8], as indicated in **Figure 3**. Some examples of DES for each type are indicated in **Table 1**.

Type I DESs are combination of quaternary ammonium salt and metal chloride. This type of DES has high melting points due to non-hydrated metal halides, making Type I DESs difficult for the isolation of lignin. Whereas, Type II DESs are combination of quaternary ammonium salt and metal chloride hydrate. The differences

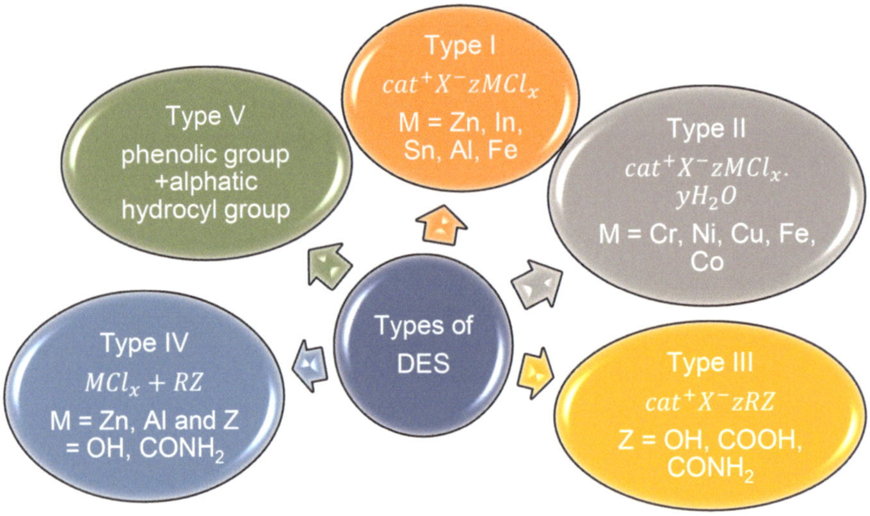

Figure 3.
Types and general chemical formula expression of DES.

Type of DES	Examples
Type I	ChCl: $FeCl_3$, ChCl: $CoCl_2$, & ...
Type II	ChCl: $CrCl_3$. $6H_2O$, ChCl: $CoCl_2$. $6H_2O$, & ...
Type III	ChCl: H_2NCONH_2, ChCl: CH_3CONH_2, & ...
NADES	ChCl: LA[a], ChCl: Gly[b], ChCl: CA[c]: H_2O, & ...
Type IV	$ZnCl_2$: CH_3CONH_2, $ZnCl_2$: H_2NCONH_2, & ...
Type V	Phenanthroline: thymol, & ...

[a]*Lactic acid.*[b]*Glycerol.*[c]*Citric acid.*

Table 1.
Some commonly used deep eutectic solvent types.

between Type I DESs and Type II DESs are the type of metal halide, non-hydrated and hydrated metal halides, respectively. Thus, in biomass processing, Type I DESs have limited applications because of high melting points of non-hydrated metal halides, but Type II DESs are more practical for industrial operations due to the hydrated metal halides [8].

Type III DESs are combination of quaternary ammonium salt and hydrogen bond donor (HBD). The HBDs can be various types of carboxylic acids and this type of DES is mostly chosen for biomass pretreatment. Type III eutectics are the most studied types of DESs, which are typically synthesized from choline chloride (ChCl) and different hydrogen bond donors [15]. Choline chloride (ChCl), a quaternary ammonium salt, and amides, alcohols, and carboxylic acids are common hydrogen bond acceptors because of their low cost, high biodegradability, low toxicity, and bioavailability [11]. The DES could also be synthesized from a molecule that is typically safe and accessible, such as urea, organic carboxylic acids, and polyols [11]. With an ever-growing library of constituents estimated at 10^6–10^8 possible binary combinations of Type III DESs have been organized into subcategories based on their behavior or nature, which include hydrophobic DES [16], metal-free DES [17], and natural DESs [18].

Type IV DESs are combination of metal chloride and HBD such as urea and ethylene glycol [11, 12]. Lately, new classes of DESs made entirely of non-ionic, molecular hydrogen bond acceptors and hydrogen bond donors have been proposed as a novel DES type, which is Type V DES with the distinctive low melting point caused by strong hydrogen bonding [19]. In addition, natural deep eutectic solvent (NADES) type of DES was also developed, which was most likely similar to two type (III) DESs, the only difference being the solvents in NADES are synthesized from natural biomass (e.g., glycerol).

2.2 Potentials of deep eutectic solvent (DES)

One of the primary reasons for the growing research interest in deep eutectic solvents is their physicochemical properties, such as; low volatility, nonflammability, low vapor pressure, easy to prepare, biodegradability, and stability [7]. In addition, DESs are so tunable, which can be designed to form a number of different types of DESs. Understanding the fundamental properties of deep eutectic solvents is the crucial thing to prepare deep eutectic solvents and to use for different applications. Thus, the properties of DES are solvatochromic, physicochemical, electrochemical, and biological properties. The common characteristics of deep eutectic solvents are freezing point, melting point, density, viscosity, conductivity, surface tension, molar heat capacity, compensability, refractive index, polarity, and dielectric conductivity which are described in this book chapter [20]. Whereas some properties of DES such as toxicity, sustainability, and biodegradability were not considered since they were considered chemically green solvents [21].

2.2.1 Solvatochromic properties

Solvatochromic properties are the measure of DES effectiveness for fractionation of lignin using empirical parameter known as Kamlet-Taft (K-T) [4]. Kamlet-Taft parametric model is the quantitative tool that determines the properties of DES based on the parameters, such as hydrogen bond donating ability, hydrogen bond accepting ability, and bipolarity of DES; represented by α, β, and π^*, respectively. The properties of these parameters depend on the nature of hydrogen bond donor (HBD) and hydrogen bond acceptor (HBA) [3]. Thus, to better understand the lignin solubility in DES, all the Kamlet-Taft parameters (α, β, and π^*) have to be determined carefully. The higher values of α and β mean high hydrogen bond donating ability and high lignin fractionation ability [22]. Similarly, high values of β and π^* dissolve high amount of lignin [23]. In addition, the solubility of lignin in DES also depends on bipolarity value (π^*) and type of lignin [24].

In addition, Hildebrand and Hansen solubility parameters are also used to determine the solubility of lignin in DES using molecular dynamics (MD) simulation [25]. It is computed based on the projected infinite dissolution activity coefficient by molecular dynamics simulation, which assists in the selection of suitable DES for lignin isolation. Nowadays, the interaction between lignin and DESs, regulated by specific characteristics, is still underexplored. However, theories suggest that DESs can be modified by altering the variety of eutectic solvents. This could lead to a more efficient and desirable DES pretreatment system, allowing for the fractionation of lignin with desired qualities for subsequent valorization.

2.2.2 Physical, chemical, electrical, and biological properties of deep eutectic solvent

2.2.2.1 Freezing point

Freezing point, a temperature at which the liquid becomes solid at atmospheric pressure, is a determining parameter of DES. The freezing point of DES lies between −79 and 149°C [7]. Though, at the freezing point below 50°C solvents are more attractive, since it's safe to use them and they are also inexpensive [26]. Factors, such as the type of ionic symmetry [8], alkaline chain length, strength of hydrogen bond [27], reaction temperature [28], molar ratio of hydrogen bond acceptor to hydrogen bond donor [29], the nature of hydrogen bond acceptor, the nature of hydrogen bond donor, and the nature of organic salt and its anion, affect the values of freezing point of the mixture [26]. In general, to prepare DES solutions for different applications, DESs with low freezing points are preferred. Low freezing points broaden the operating temperature range and facilitate the transmission of intermediary substances during separation processes [30]. For example, DESs with low freezing points are preferred for biomass processing applications, as they can remain as liquids for better mixing and mass transfer efficiency [31]. The freezing points of HBA and HBD are essential reference properties when designing DESs, as indicated in **Table 2**. The lower the freezing points of HBA and HBD of DES components, the deeper the decrease in the prepared DES.

2.2.2.2 Density

Density is likely the most crucial property to identify the molecular mobility and activity of DES, which varies with the type of eutectic mixture. Most research studies showed that the density of deep eutectic solvents is greater than that of water, which is in the range of 1.0–1.3 g.cm^{-3} at atmospheric temperature. Whereas the densities of DESs obtained from metallic salts like zinc chloride ($ZnCl_2$) are between

HBA	Mp(°C)	HBD	mp(°C)	Molar ratio	DES T_f(°C)	Ref.
ChCl	303	Urea	134	1:2	12	[32]
ChCl	303	Thiourea	175	1:2	69	[29]
ChCl	303	1,3-dimethyl urea	93	1:2	29	[29]
ChCl	303	Citric acid	149	1:1	69	[32]
ChCl	303	Adipic acid	153	1:1	95	[32]
ChCl	303	Malonic acid	134	1:1	10	[32]
ChCl	303	Oxalic acid	190	1:1	34	[32]
$ZnCl_2$	293	Urea	134	—	9	[12]
$ZnCl_2$	293	Acetamide	81	—	−16	[12]

Table 2.
A comparison of freezing point values of different deep eutectic solvents.

DES (HBA: HBD)	Molar ratio	T(K)	Density (g/cm^3)	Ref.
ChCl: urea	1:2	313.15	1.24	[36]
		298.25	1.212	[36]
MTPB: GLY	1:3	363	1.2513	[34]
BTPC: GLY	1:16	363	1.1968	[34]
ATBP: GLY	1:14	363	1.2201	[34]
ChCl: GLY	1:2	363	1.1548	[34]
DAC: GLY	1:2	363	1.1342	[34]
TABB: GLY	1:4	363	1.1064	[34]
ChCl: D-glu	1:1	358	1.2678	[35]

Table 3.
Densities of different deep eutectic solvents.

1.3 and 1.6 g.cm^{-3} [26]. In general, type I-IV DESs have higher density than water, whereas the density of type V DES is lower than that of water [9]. In addition, the density of hydrophobic deep eutectic solvent is lower than that of water changing between 0.889 and 0.93 based on the components [14].

The density of DES depends on the choice of the hydrogen bond donor and the molar ratio [33]. In addition, the density of deep eutectic solvent is temperature-dependent, which decreases linearly with increasing temperature [33]. Alomar et al. investigated the density of DES from glycerol. In the study, six different DESs were prepared taking glycerol as a HBD and six different salts, i.e., methyltriphenyl-phosphonium bromide (MTPB); benzyltriphenylphosphonium chloride (BTPB); allyl-triphenyl-phosphonium bromide (ATPB); choline chloride (ChCl); N, N diethylethanolammonium chloride (DAC); and tetra-n-butylammonium bromide (TBAB).

The measured density in the temperature ranges of 298–363 K showed the highest density 1.2965 g/cm^{-3} at 298 K [34]. Whereas, Hayyan et al. characterized physiochemical properties such as density, viscosity, refractive index, surface tension, and pH of glucose-based deep eutectic solvents. The results revealed that DES densities are above 1.20 g/cm^3. In general, we can say the density of DES decreases as the temperature increases, due to the increased activity and molecular mobility [35]. However, the density of DES may also vary with the type of DES component and molar ratio, as shown in **Table 3**. As a result, the density of DES should be optimized to fit lignin fractionation.

2.2.2.3 Viscosity

One of the essential and well-studied characteristics of DES is viscosity. A number of researchers have described that the viscosity of DES is being very viscous at room temperature (> 100 mPa.s.). The reason behind the high viscosity of DES can be due to the hydrogen bond network that exists between the components of DES, as indicated in **Table 4**. However, deep eutectic solvents may present a broad viscosity range from low viscosity to very high viscosity. For instance, choline chloride: ethylene

DES (HBA: HBD)	Molar ratios	T(K)	Viscosity (mpa s)	Ref.
ChCl: urea	1:2	303.15	152,449	[29, 37]
ChCl: GLY	1:2	303.15	190,246.79	[38, 39]
ChCl: EG	1:2	303.15	35	[40]
ChCl: EG	1:2	298.15	37	[41]
ChCl: MA	1:1	303.15	982.9,464,608	[42, 43]
ChCl: OA	1:1	313.15	1881,2142,202	[42, 43]
ChCl: LV	1:2	303.15	164.5	[42]

Table 4.
Different types of deep eutectic solvents with their viscosities.

glycol has low viscosity, whereas sugar-based DESs have very high viscosity (12,730 mPa.s for 1:1 ChCl: Sr. (sorbitol) at 30°C and 34,400 mPa.s for 1:1 ChCl: glu (glucose) at 50°C), and metal salts-based deep eutectic solvents have even higher viscosities (85,000 mPa.s for 1:2 choline chloride: zinc chloride at 25°C) [26].

Factors, such as electrostatic interactions, chemical composition of the DES component, reaction temperature, and amount of water, all this affect the viscosity of deep eutectic solvent [7]. In addition, the experimental methods, sample preparation, and the impurities also affect the viscosity of deep eutectic solvent [42]. DES with high viscosity is not favorable for lignin fractionation. Thus, the development of DESs with low viscosities is particularly sought for lignin isolation. Using the hole theory model, it's possible to prepare low viscosity deep eutectic solvents, in addition to using fluorinated HBD and small cations.

2.2.2.4 Conductivity

Conductivity is the measure of DES strength for conducting electricity. The value of DES conductivity depends on the viscosity of DES and other properties of DES. The hydrogen bond acceptor to hydrogen bond donor molar ratio, the type of both the organic salt and the hydrogen bond donor, as well as the anion of the salt [36], and the imbibition of water are the factors that affect conductivity of deep eutectic solvents [44]. In addition, the quantity of charged compounds and their mobilities influence DES conductivity. For instance, the conductivity of a deep eutectic solvent is inversely proportional to the viscosity of DES. Since deep eutectic solvents are mostly viscous, it implies low ionic conductivity, which is a problem for electrochemical applications [45]. As a result, raising the temperature reduces viscosity while increasing conductivity [45]. Generally, DESs display lower conductivity that is much lower than 2 ms cm^{-1}.

Thus, microwave irradiation has recently been widely used to enhance lignin fractionation. When microwave heating is used, DESs with high electrical conductivity and dielectric constant effectively delignify lignocellulosic biomass by molecular collisions. Furthermore, for electrochemical transformation of lignin in adequate DES conductivity is required. The electrical conductivity of DESs can be tailored by optimizing the DES constituents and their molar ratios.

HBA: HBD	Molar ratio	T(°C)	Surface tension (mN/m)	Ref.
ChCl: urea	1:2	298.15	52.0	[47]
ChCl: glycerol	1:2	298.15	56.0	[47]
ChCl: EG	1:2	298.15	48	[47]
ChCl: MA	1:1	298.15	65.7	[47]
TPAB: glycerol	1:3	303.15	46	[48]
TPAB:EG	1:3	303.15	40.1	[48]
TPAB: tri EG	1:3	303.15	39.3	[49]

Table 5.
Surface tension for selected deep eutectic solvents.

2.2.2.5 *Surface tension*

Surface tension is a measure of the cohesive forces in a liquid on its surface and a measure of the inherent energy between liquid molecules. It is a critical physical quality for the determination of a variety of applications, such as wettability, mass transfer capacity for liquid-liquid or gas-liquid extraction, and multiphase homogeneous catalysis. Surface tension is greatly influenced by the intensity of intermolecular interactions between the HBDs and the appropriate salt. For most of the examined systems, the surface tension of DES is exceptionally high. The values relative to the reported deep eutectic solvents generally vary between 35 and 75 mN/m at 25°C [36].

The surface tension of deep eutectic solvents is affected by the extra hydroxyl group or a longer alkyl chain in the quaternary ammonium salt, the salt mole fraction, cation type, and temperature [31]. The surface tension increases linearly with decrease in temperature [46] and it also decreases with increasing salt mole fraction due to the weakening of the hydrogen bond donor bonding network during mixing [36]. To understand further the variation of surface tension with different types of DESs, molar ratios at different temperatures are indicated in **Table 5** below.

2.2.2.6 *Polarity*

Polarity is an important property for determining the solubility of a deep eutectic solvent measured by solvatochromic parameters [50]. The most commonly used polarity scales of Reichardt and Dimroth's are normalized ET scale [50] and the multiparameter scale of Kamlet-Taft (α, β, and π^*) [51, 52]. However, Abbott and colleagues reported the first attempts to investigate the solvation processes in deep eutectic solvents for eutectic mixes of choline chloride and glycerol.

Abbott et al. characterized the solvent polarities of ChCl and glycerol at different molar ratios using the indicated methods. The measured polarity of DES by ET [29] polarity scale showed that ChCl and glycerol DESs are polar fluids. These polarity scale ET [29] values of ChCl and glycerol are comparable to those of known polar HBD solvents. The reported result showed the change in polarity scale ET [29] values with increasing ChCl from 57 kcal/mol for pure glycerol to 59 kcal/mol for pure ChCl (extrapolated) with an almost linear trend [53]. Though, it's understood that to compare polarity parameters obtained by different solvatochromic probes is so difficult as the polarity scales are not universal and are probe dependent [54–56].

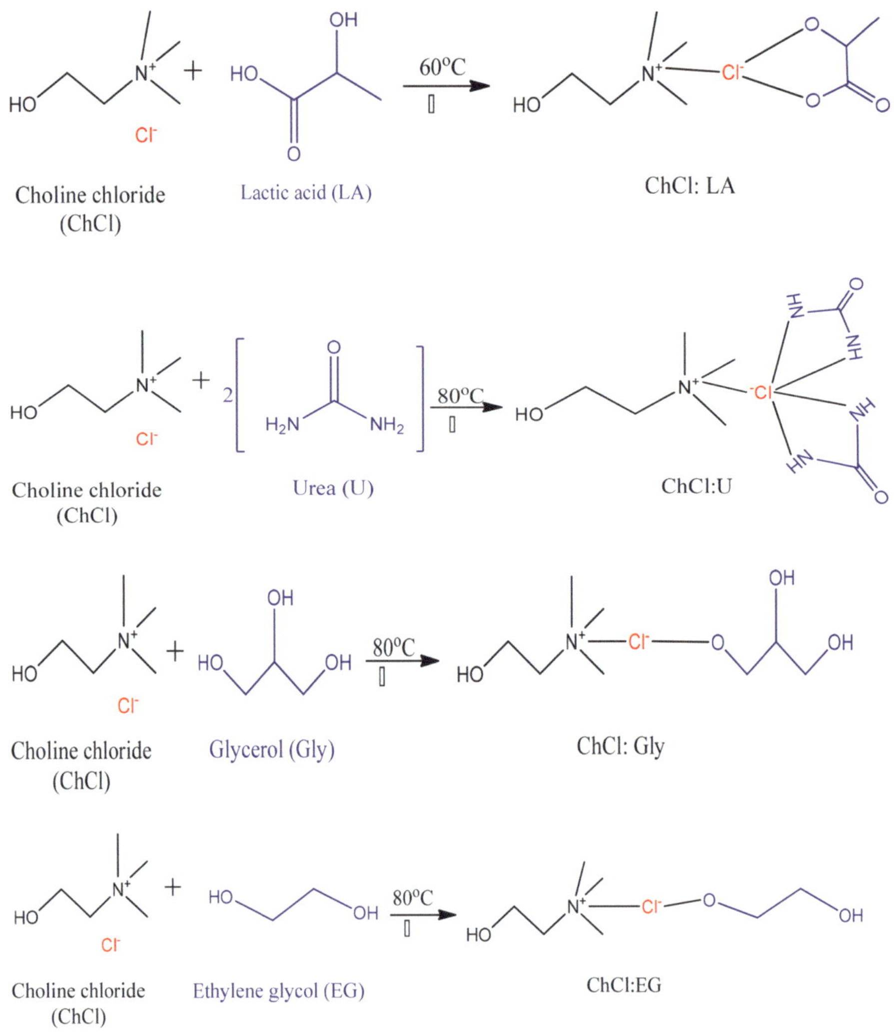

Figure 4.
Chemical reaction followed for the synthesis of DES from commonly used type 3, HBA and HBD as an example.

3. Synthesis and characterization of deep eutectic solvent

See **Figure 4**.

3.1 Deep eutectic solvent synthesis

Deep eutectic solvent (DES) is synthesized by simply reacting hydrogen bond donor (HBD) and hydrogen bond acceptor (HBA) at specified molar ratios, as shown in **Figure 4**. During the preparation of DES, factors such as the water content and purity of individual components of HBA and HBD, as well as the storage and drying time of the prepared DES, should be consistently considered. The presence of these factors leads to variations in the properties of DES and preparation. Since most components of HBA and HBD absorb moisture in the atmosphere and disrupt the physiochemical properties of DES, furthermore, it is important to note that reagents and DES should be stored on moisture-free surfaces [57].

Several methods were developed for the preparation of DES, as shown in **Table 4**, to ensure their reproducibility and constant property. The evaporation and grinding are the most commonly employed extraction methods [7, 42]. However, recently advanced extraction methods, including microwave-assisted approach [58], ultrasound-assisted, and freeze-drying synthesis methods, were developed considering extraction time and energy consumption during extraction process [18, 20].

The preparation of DES by heating and stirring methods is carried out by mixing and heating the compounds at a temperature of 50 to 100°C under constant stirring until a homogeneous liquid is formed, as shown in **Figure 5**. Following the preparation, the colorless eutectic solvent will be formed, which is in this case DES, and finally it will be dried to remove extra moisture [42]. This method is the simplest and most common one. However, there is a limitation, irrespective of the production process: a high temperature may result in the deterioration of the DES due to the esterification reaction of solvents based on choline chloride and carboxylic acids [59]. In addition, uncontrolled stirring speed and heating temperature make the eutectic mixture crystallize. Thus, optimizing the appropriate value for reaction time and heating temperature is required to use the heating and stirring methods.

The evaporation method is another way of deep eutectic solvents' preparation conducted by dissolving DES components in water, followed by evaporation at 50°C and then dropping the solution in a desiccator [60]. The addition of water is performed to enhance the mixing of highly viscose sugar-based DES. Compared to the heating and stirring methods, the evaporation method consumes less power and is mostly suitable for the preparation of NADES, which commonly consist of amino acids, organic acids, and sugars [60].

The grinding method is also the commonly used deep eutectic solvent preparation method, executed by mixing the eutectic solvents at room temperature, then grinding them together using mortar and pestle, until a clear liquid is seen [42]. Due to its

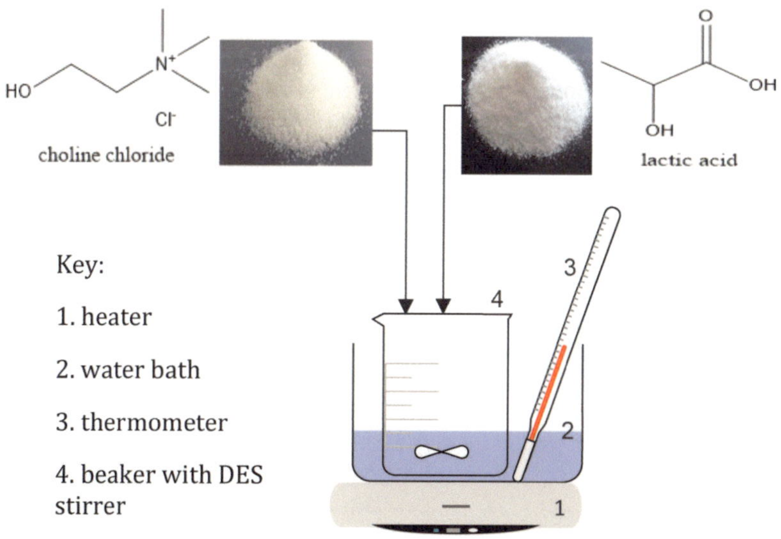

Figure 5.
Schematic diagram for the synthesis of DES by the heating method.

hygroscopic nature, the deep eutectic solvent absorbs water, so it has to be stored in a vacuum for hours. In comparison to the heating and evaporation methods, grinding may be done at a lower temperature. However, in order to use the evaporation process, HBA and HBD must be water-soluble. Furthermore, total water removal might be time-consuming [61].

Freeze-drying preparation of deep eutectic solvent is carried out by adding a stoichiometric proportion of HBA and HBD, and then diluting with distilled water to obtain a 5% aqueous solution. To create a clear, viscous liquid, the aqueous solution is frozen at a very low temperature (253 K) and then freeze-dried at a lower temperature [62]. Gutiérrez et al. examined the freeze-drying approach for DES production. The researcher noticed that high-purity DESs may be synthesized and that they are especially ideal for the creation of DESs containing unstable thermal chemicals [63]. However, in terms of energy cost, time cost, and economic sustainability, it is not suited for large-scale manufacture. Thus, searching for advanced methodologies that fill the limitations is crucial for the reliability of DES applications.

Microwave-assisted DES preparation is one of the fastest, cheapest, easiest, and eco-friendly methods [58]. The principle behind microwave-assisted DES preparation is executed by mixing eutectic solvents and placing in irradiation for 20 s after enclosing the solvents in a 20-ml cylinder. Kohli et al. studied microwave-assisted DES preparation, which was used to improve the fractionation of lignocellulosic biomass to yield lignin for biochemical synthesis [21].

Ultrasound-assisted preparation is another recent development for deep eutectic solvent preparation. The process is similar to microwave-assisted preparation of DES, which is performed by mixing HBA and HBD in a glass cylinder stoichiometrically [64]. Then put the mixture in an ultrasonic bath for 1 to 5 h. The ultrasonic heating temperature can be increased from room temperature to 60°C, depending upon the DESs constituents. Finally, stable and non-crystallizable DES can be obtained, after keeping the solution for 24 h in the same container under ambient conditions to ensure the formation of a homogeneous mixture [64].

Espino et al. reviewed different preparation techniques of DES. All the above synthesis techniques and others have their own pros and cons; the thing is, it's critical to know the approaches of DES preparation after utilizing for different applications. Since, DESs are prone to degradation and the formation of byproducts is based on the synthesis conditions. For instance, DES degrades when synthesized at high temperatures (**Table 6**) [62].

Methods	Compositions	Findings	Ref.
Heating and stirring at 80°C	ChCl: LA (1:2) 20% dilution	Phenolic extraction from *Cajanus cajan* leaves	[62]
Heating and stirring at 100°C	ChCl: LA (1:2) 20 v/v% dilution	Saponin extraction from *Agave sisalana*	[62]
Evaporation method at 50°C		Preparation of DES evaporation methods operated at low-energy heating method and it is mostly suitable for NADES preparation	[65]
Ultrasonication (1 to 5 h at 60°C)	75 v/v% dilution Different alcohol-based DES	Extraction of gingerols from ginger. Extraction T > 50 °C induced degradation and loss of phenolic compounds. DESs are stable and not prone to crystallization	[64]

Methods	Compositions	Findings	Ref.
Heating and stirring at 60°C, 200 rpm Microwave for enhancement.	ChCl as HBA with six different carboxylic acids as HBD	Delignification of lignocellulosic biomass	[21]
Heating and stirring methods at 100°C Grinding method at room temperature	ChCl as HBA with carboxylic acid	To assess the properties of DES (densities, viscosities, and refractive index). The case of viscosity grinding methods preparation is preferred than heating method.	[42]
Microwave-assisted method. For 200 W to 800 W and 5 to 30 s.	Three different NADES prepared	To develop. Microwave-assisted method is an effective NADES preparation method. Since it helps to prepare NADES in short period of time and reduces conventional preparation time from tens of minutes (30–90min) to 20 sec, and the energy consumption, 650 times less. 200 W for 20 sec is the best value to reach eutectic point.	[58]

Table 6.
A summary table of deep eutectic solvent synthesis methods.

3.2 Characterization of DES physicochemical properties

The physical and chemical properties of deep eutectic solvent have to be determined, prior to using deep eutectic solvents for different applications, including the isolation of lignin from lignocellulosic biomass [66]. Thus, physical and chemical properties that affect the DES properties and are to be characterized include phase behavior, density, viscosity, ionic conductivity, surface tension, and polarity.

Characterization tools, such as ultraviolet-visible (UV-vis) spectroscopy, Raman, infrared (IR), and fluorescence correlation spectroscopy *(FCS) and others, can be used to characterize and understand the molecular dynamics of DESs [67]. In addition, thermogravimetric analysis (TGA) for determining thermal properties, Fourier transform infrared spectroscopy (FTIR) for determining the functional groups of DES, pycnometer for performing density analysis, and viscometer for conducting viscosity analysis [44, 59], as shown in **Figure 6**.

For instance, Rodriguez Rodriguez et al. characterized biodegradability of choline chloride and carboxylic acid-based deep eutectic solvents prepared by the heating method. The researcher studied the degradation of DES using TGA and nuclear magnetic resonance (NMR) techniques. The reported result showed that the DES of this family is prone to degradation between 125 and 235°C due to esterification reaction and from the ^{1}H NMR (proton nuclear magnetic resonance) analysis, it was also confirmed that the DESs had esterified to a certain extent [59].

Dai et al. characterized the physiochemical properties such as viscosity, density, conductivity, and polarity of NADES prepared from different chemicals using the heating method with stirring at 70°C, until a clear liquid was formed. The researcher used FTIR (Bruker FT-IR spectrometer at 25°C), NMR (40°C on a 500 MHz Bruker DMX-500 spectrometer), viscometer, coulometer (Metrohm), and density meter (DMA 5000) for characterizing prepared NADES. Furthermore, a polarity test was

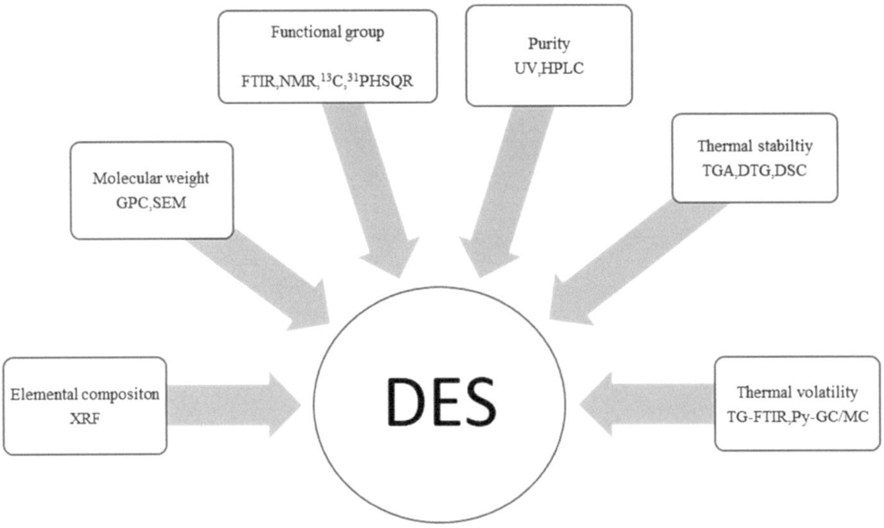

Figure 6.
Characterizing tools for determining the properties of deep eutectic solvents.

performed with Nile Red (NR) as a solvatochromatic probe to estimate the polarity of NADES in the form of molar transition energy [44].

Cui et al. prepared deep eutectic solvents by mixing choline chloride with different organic acids as hydrogen bond donors using different methods such as stirring, grinding, and stirring with grinding methods. Physical parameters (density, electrical conductivity, dynamic viscosity, and refractive index) were examined at temperatures ranging from 15.15 to 65.15°C at atmospheric pressure [33].

Alvarez-Vasco et al. characterized the lignin isolated by DES for their molecular weight and their cleavage of ether bond using GPC (gel permeation chromatography) and C NMR (Carbon-13 nuclear magnetic resonance). The NMR analysis showed the absence of the ether linkage β-O-4 after pretreatment of the milled wood. Whereas the GPC analysis showed that the molecular weight of about 890 g/mol provides evidence that DES depolymerizes native lignin by selective cleavage of ether bonds without subsequent condensation reactions [23].

4. DES mediated lignin isolation and recent advancements

Pretreatment processes are primarily concerned with the isolation of interconnected fractions of lignocellulosic contents and increasing the accessibility of each individual content, making them an essential step in a wide range of valorizations [68]. Thus, to make greater use of lignocellulosic biomass for lignin-based coproducts, a method for fractionating lignin from the feedstock with high effectiveness and low cost must be developed. Currently, physical, chemical, physiochemical, and biological methods are the major approaches developed for lignocellulosic biomass pretreatment strategies [69].

Physical pretreatment methods are commonly used to reduce biomass particle size and disintegrate lignocellulosic components. Physical pretreatments, such as milling, extrusion [70], microwave [68], and ultrasound pretreatment [71], are commonly used, and to facilitate the process, chemicals and enzymes are also used. Compared with physical pretreatments, a microwave pretreatment has several advantages, such as: short reaction times, rapid and uniform heating, low reaction activation energy, high product yield, minimal side products, cost-effectiveness, eco-friendliness, and energy efficiency [68].

Chemical pretreatment of lignocellulosic biomass, on the other hand, is performed by using either acid, alkali, ionic liquids, or combinations of one another to alter the physical and chemical structure of lignocellulosic biomass. Many chemical pretreatment techniques were studied for valorization of lignocellulosic biomass, some of which are alkali [70], acidic [72], organosolv [73], cosolvent enhanced lignocellulosic fractionation (CSELF) [74], ammonia fiber explosion (AFEX), and deep eutectic solvent (DES) pretreatment [75]. However, DES is a recently developed green solvent that is used for delignification. Properties, such as low volatility, high solubility, biodegradability, cheap cost, environmental friendliness, and structural feasibility, promoted their widespread use in the effective delignification of various biomass feedstocks [9].

For instance, Malaeke et al. solubilized lignin from wheat straw up to 50% w/w in a DES consisting of resorcinol-ChCl coupled with ultrasonic irradiation, which improves the mass transfer and dissolution process. DES is prepared by the heating method with subsequent heating at 100°C and stirring until a clear liquid is formed. During pretreatment, the solubility of cellulose in the prepared DES is very low; thus, it enhances the isolation of lignin [76].

The primary step toward lignin valorization via various techniques is effective lignin isolation from lignocellulosic biomass. Many research works on DES pretreatment have indicated easy lignin isolation/delignification using various DESs, as shown in **Table** 7. The HBA of most DESs used for pretreatment is choline chloride

DES	Ratio	Biomass	Parameters	Lignin yield	Ref.
ChCl: OA	1:1	Corncob	90°C, 24 h	98.5	[77]
ChCl: LA	1:2	Corncob	90°C, 24 h	64.7	[77]
ChCl: LA	1:15	Corncob	90°C, 24 h	93.1	[77]
ChCl: OA	1:1	Poplar	Microwave, 3 min	80	[66]
ChCl: LA	1:2	Switchgrass	Microwave,45 s	72	[78]
ChCl: LA	1:2	Corn stover	Microwave,45 s	79.6	[78]
ChCl: LA	1:2	Miscanthus	Microwave, 45 s	65.2	[78]
ChCl: LA	1:2	Poplar	145°C, 6 h	78.5	[23]
ChCl: LA	1:5	Rice straw	60°C, 12 h	60	[67]
ChCl: LA	1:10	Eucalyptus	110°C, 6 h 80	80	[79]
ChCl: LA	1:10	Willow	120°C, 12 h	91.8	[80]
ChCl: LA	1:9	Poplar	120°C, 6 h	98.1	[81]
ChCl: LA	1:5	Bagasse	80°C, 12 h	50.6	[31]
ChCl: FA	1:2	Corn stover	130°C, 2 h	23.8	[82]
ChCl: EG	1:2	Corncob	90°C, 24 h	87.6	[77]
ChCl: GLY	1:2	Corncob	150°C, 15 h	59.0	[83]
ChCl: GLY	1:2	Corn stover	180°C, 2 h	−60	[82]

DES	Ratio	Biomass	Parameters	Lignin yield	Ref.
ChCl: GLY	1:2	Switchgrass	110°C,1 h	76.6	[84]
ChCl: GLY: $AlCl_3{\cdot}6H_2O$	1:2:0.28	Poplar	120°C,4 h	95.5	[84]
ChCl: EG	1:2	Switchgrass	130°C, 30 min	78.9	[85]
GH:EG: PTSA	1:1.94:0.06	Switch grass	120°C, 6 min	82.1	[81]
ChCl: ME	1:6	Wheat straw	90°C, 12 h	81.0	[85]
ChCl: DE	1:8	Wheat straw	90°C, 12 h	73.5	[85]
ChCl: imidazole	3:7	Corncob	115°C, 15 h	76.7	[83]
ChCl: Urea	1:2	Corncob	80°C, 15 h	27.1	[83]
ChCl: Urea	1:2	Rice straw	130°C, 8 h	44.7	[86]
ChCl: PCA	1:1	Switchgrass	160°C, 3 h	60.8	[87]
TEBAC:LA	1:9	Wheat straw	100°C, 10 h	79.7	[88]
ChCl: LA	1:5	Oil palm empty fruit bunch	120°C for 8 h	88.0	[89]
ChCl: OA	1:2	Rice straw	120°C for 1 h	75.9	[90]
BTMAC:LA	1:2	Corncob	140°C, 2 h	63.4	[91]

Table 7.
Analysis of lignin isolation using DES from different lignocellulosic biomasses.

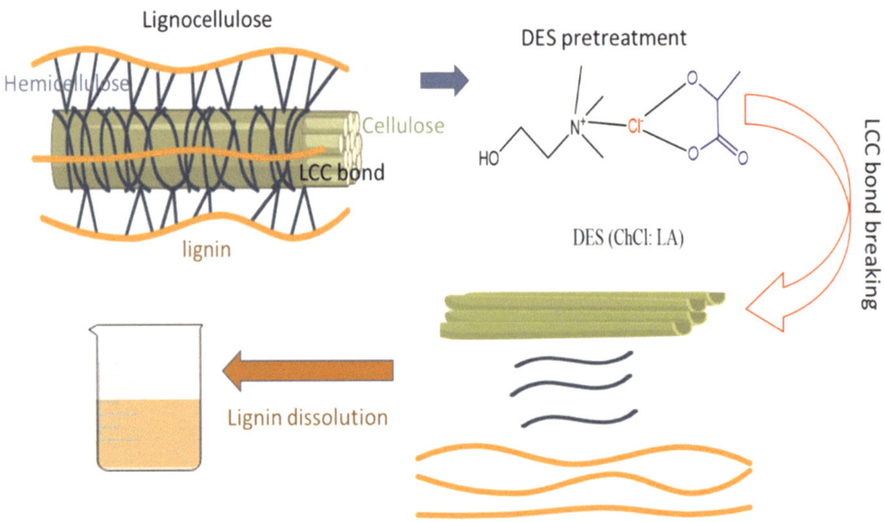

Figure 7.
General schematic diagram of lignin isolation using DES.

(ChCl), whereas the HBDs are more diverse, with carboxylic acids, polyols, and amides often being utilized. Several carboxylic acids generated from biomass, like formic acid (FA), lactic acid (LA), and oxalic acid (OA), have been used as HBDs in DES synthesis.

Deep eutectic solvents (DESs) from choline chloride as HBA and carboxylic acid, polyols, and others as HBDs were used for the isolation of lignin from lignocellulosic biomass. From others, DES from choline chloride as HBA with carboxylic acid as HBD solvent are the most effective for lignin isolation. However, the process takes long time at an operating temperature of 60 to 150°C, which needs modification. Recently, high-performance microwave-assisted DES treatment is introduced, intended to reduce time and energy [66, 77].

Deep eutectic solvent (DES) from choline chloride as HBA and polyols that include glycerol, ethylene glycol, and propylene glycol as HBD is another type of DES that was used previously for lignocellulose pretreatment. From this category, ChCl: GLY (choline chloride: glycerol) is an effective polyols-based DES-lignin isolation, which yields 60–90% [66, 83]. Likewise, it has limitations in reaction time and operating temperature as the previous one. In order to overcome these limitations, acidified polyol was introduced [84]. In addition, to further shorten time and enhance yield, ternary DES was developed [81, 92]. The schematic representation of the general mechanisms of lignin isolation using DES was illustrated in **Figure 7**.

5. Factors affecting DES-lignin pretreatment

The key process parameters that distinguish DES pretreatment of lignocellulosic biomass from others are reaction temperature, reaction time, molar ratio, and type of DES component [93]. Hong et al. categorized the parameters that affect DES pretreatment efficiency as characteristics of biomass (components of raw biomass, crystallinity of biomass, and particle size), characteristics of DES (HBA, HBD, and molar ratio), and reaction conditions (reaction time, temperature, and solid-to-liquid ratio) [10].

For instance, Okur et al. studied the effects of DES component molar ratio, pretreatment temperature, and pretreatment time on lignin removal effectiveness. The raw material paddy husks and the DES solvent (ethylene glycol) EG: CA (citric acid) were utilized in the investigation. The study reported varied lignin yields in percentage (47.10, 52.35, and 51.17%) utilizing DES of EG:CA molar ratios of 1:1,1:2, and 2:1 at 90°C. It has been found that pretreatment temperature and reaction time have a substantial impact on the production of lignin from paddy husks. Because intermolecular hydrogen bonds are decreased at high temperatures, the viscosity of the DES solution decreases, DES rapidly enters the lignocellulosic structure, and therefore the lignin removal efficiency rises. In this work, the most significant parameters that affect deep eutectic solvent pretreatment of lignocellulosic biomass, such as temperature, molar ratio, reaction time, types of biomasses, and types of DES solvents, are included.

5.1 Reaction temperature

The pretreatment temperature is a critical key parameter influencing the degree of delignification in the DES pretreatment response [93]. For instance, increasing temperature decreases viscosity of DES [21]. In addition, DES may enhance the breaking of huge quantities of ether bonding, particularly β-O-4 bond, and the resulting lignin becomes lower molecular weight lignin. Thus, the removal of lignin is highly impacted by reaction temperature [6]. In general, the commonly used DES pretreatment temperature ranges from 60 to 200°C [94, 95].

Y. Liu et al. studied the influence of temperature on lignin isolation by DES (the molar ratio of ChCl to p-coumaric acid was 2:1) for 7 h with a solid-to-liquid ratio of 1:10. When the pretreatment temperature was increased from 140 to 160°C and 180–200°C, consequently, delignification rate was increased from 8, 10, and 50 to 84%, respectively, and similarly, for the second temperature range, delignification rate was changed from 27, 36, and 66–90%, respectively. Ma et al. also performed experiments in order to improve wheat straw fractionation efficiency using a deep eutectic solvent (DES) solution comprised of chlorine and lactic acid. In the study, wheat straw pretreatment was conducted by utilizing chloride: lactic acid (1:2) and DES solution (1:20 solid-to-liquid ratio) at various temperatures (90, 120, 150, and 180°C) for 6 and 12 h. It is reported that the yields of lignin were increased from 15.26 to 81.54% and purity of 91.33% with temperature increase from 90 to 180°C. Thus, the optimal temperature obtained in the pretreatment of wheat straw for lignin isolation was 150° C for 6 h [96].

Lou et al. conducted a comprehensive analysis of lignin nanoparticle synthesis from wheat straw using ChCl: LA-based DES. In the study, process parameters, such as temperature (90 to 150°C) and reaction time (6 to 24 h), were considered. The reported result showed high-purity lignin (up to 94.8%) with a high yield (up to 81.5% from air-dried samples and 85.9% from oven-dried samples) at 150°C for 24 hours [97]. This indicates that ChCl: LA DES can isolate lignin efficiently at optimum reaction time and temperature.

C. W. Zhang et al. investigated the effects of different DES combinations, pretreatment temperature, and reaction time on the utilization of DESs in the pretreatment of lignocellulosic biomass. Analyzing the effects of pretreatment temperature on lactic acid with ChCl-treated corncob (2:1) resulted in that lignin extractability improved (18.1–95.5%) when the pretreatment temperature goes from 70 to 110°C. In the study, it was concluded that the reaction time of 24 h and temperature of 90°C were the optimum values obtained. It can be concluded that pretreatment temperature is one of the most important aspects of DES pretreatment of lignocellulosic biomass, and optimizing reaction temperature to improve lignin removal from lignocellulose biomass will be one of the research hotspots [77].

5.2 Reaction time

The pretreatment time of DES is found to be in the range of 0.5 to 24 h, based on different researches [23, 77]. As a result, optimum pretreatment time should be adequate not only to achieve a high lignin yield. Since, the solid recovery rate can be lowered by longer pretreatment time because carbohydrate components in lignocellulosic biomass could be dissolved and decomposed clearly. Thus, it is required to optimize reaction time in the pretreatment of lignocellulose to get the required yield for valorization of lignocellulosic biomass.

For instance, Liu et al. processed wheat straw at 100°C with a DES (triethyl benzyl ammonium chloride: lactic acid) molar ratio of 1:9 and lecithin: sphingomyelin (LS) ratio of 1:15. To determine the effect of time on DES pretreatment, the effects of different pretreatment times of 2, 4, 6, 8, 10, and 12 h were examined. The recovery rate of the cellulose-enriched residues rapidly dropped with increasing pretreatment duration, falling from 66% in 2 h to 48.7% in 12 h. Meanwhile, the xylan and lignin removal rates improved steadily from 43% in 2 h to 78% in 12 h and 48% in 2 h to 74% in 12 h, respectively [88]. For different molar ratios, reaction times, DES compositions, and raw materials, the increment lignin yield with reaction time was confirmed by the other researcher. Li et al. used ChCl: LA at different molar ratios of (1:2, 1:4, 1:6, 1:8, 1:10, and 1:12), temperature at (90–120°C), and reaction time (4–62 h) for delignification of willow. The result indicates that the efficiency of delignification increased significantly with increasing treatment time from 6 h to 12 h at molar ratio of (1:10). Delignification at ChCl: LA (1:10), temperature (20°C), and 12 h resulted in 91.82% lignin yield but when the treatment time extended to 42 h, the lignin yield increased to 94.2% [66]. However, as the treatment time increased, the solid residue yield decreased, indicating that the carbohydrate components dissolved and/or degraded significantly. Thus, reaction time of 12 h is the optimum time for delignification.

Zhang et al. investigated the effects of temperature (70 to 110°C) and time (1.5 to 36 h) at fixed DES molar ratio of ChCl: LA (2:1) on delignification of corn cup. The effects of pretreatment time on the corncob by lactic acid: ChCl (2:1) were studied with the elongation of pretreatment time from 1.5 to 36 h at a temperature of 70 to 110°C. The reported results showed that the lignin extractability efficiency increased significantly with time. The optimal pretreatment temperature and time were shown to be 24 h and 90°C. In general, most of the research works reported that increasing reaction time favors to increase the yield of lignin extraction. Increase the removal rate of lignin, which increased the accessible area of polysaccharides. Another significant reason was that as time goes by, better swelling of cellulose fibers could be obtained by pretreatment solvent's penetration into biomass fiber bundles [77].

5.3 Molar ratio of HBA to HBD

The molar ratio of DESs, i.e., the ratio between hydrogen bond acceptor (HBA) and hydrogen bond donor (HBD), has effect on DES formulation. Hence, different formulations of HBA to HBD were discovered. From the different types of DES combinations, ChCl: LA formulation is the most commonly researched type. Since, this combination is effective to isolate lignin [98, 99]. For willow lignocellulosic biomass, Li et al. utilized ChCl: LA at different molar ratios (1:2, 1:4, 1:6, 1:8, 1:10, and 1:12) to investigate their effect on the isolation of lignin from willow at temperatures of 90–120°C and time at 4–6 h. It was shown that choline chloride to lactic acid with a molar ratio of 1:10 was an optimum molar ratio yielding a maximum lignin amount, 91.82% [100].

For corncob, C. W. Zhang et al. conducted experiments using an LA: ChCl DES solvent and investigated the impacts of DES molar ratio (2:1 to 15:1), pretreatment temperature (70 to 110°C), and reaction time from 1.5 to 36 hours on the delignification of corn cup. Thus, the result showed that the lignin extractability increased (64.7–93.1%), as the molar ratio of lactic acid to ChCl increased from 2:1 to 15:1 [77]. For different biomasses and DESs, Okur et al. studied the effects of molar ratio, temperature, and time. In the study, rice husks were pretreated using an EG: CA binary DES system. Pretreatment tests were carried out at 90°C for 4–24 h and at 120°C for 1–16 h with EG: CA molar ratios of 1:1, 1:2, and 2:1. The percentage of delignification efficiency for DES systems with EG: CA molar ratios of 1:1, 1:2, and 2:1 was determined to be 47.10, 52.35, and 51.17%, respectively. After 4 hours of pretreatment at 120°C, the maximum lignin removal was determined to be 57.33%, which indicates that delignification efficiency with varying molar ratios was found to be relatively similar [101]. Thus, process parameters, such as temperature and time, were considered to be more critical parameters influencing the delignification of rice husk. To confirm, a detailed analysis using different DES formulations is required.

Tan et al. synthesized five distinct DESs (ChCl: LA, ChCl: MA, ChCl: CA, ChCl: FA, and ChCl: AA) to investigate the effects of molar ratio on the delignification of lignocellulosic biomass (empty oil palm fruit bunch). The results indicated that increasing HBD concentration enhanced lignin extraction by 61% and 40% in the ChCl: LA and ChCl: AA pretreatments, respectively, whereas lignin extraction was only 6% improved in the ChCl: MA pretreatment. When the ratio of ChCl to LA was adjusted from 1:1 to 1:15, lignin production gradually increased from 33 to 61%. On the other hand, increasing the molar ratio from 1:2 to 1:5 resulted in a significant decrease in lignin output from 62 to 25%. This implies that a 1:5 ratio may not be an appropriate eutectic composition for cycloserine-cefoxitin fructose agar (CC-FA) DES and for delignification. It can be observed that high amount of lignin yield (61 and 62%) was obtained using two different DESs at optimum molar ratios; ChCl: LA at 1:15 and ChCl: FA at 1:2, respectively [98, 99]. Hence, optimizing molar ratio of DES for better isolation of lignin is worth.

5.4 Type of DES component

Acidic DES is commonly used in the pretreatment of lignocellulosic biomass because it effectively delignifies lignin and hemicellulose and significantly improves the enzymatic hydrolysis of lignocellulosic pretreatment residues [77]. The DES generated from ChCl: Urea generally fractionates lignin by cleaving the ether bond. Under alkaline circumstances, the breakdown of the β-O-4 bond in non-phenolic-type lignin model molecules can progressively generate epoxide intermediates. In the presence of alkaline DES, the phenolic hydroxyl group may produce quinone methylator-type intermediates due to deprotonation, which then undergoes depolymerization and repolymerization.

C. Li et al. investigated the effects of acidic DES on lignocellulose pretreatment. In the study, sugarcane bagasse and five DESs: ChCl: Urea, ChCl: EG, ChCl: Gly, ChCl: LA, and ChCl: OA were used. It reported that acidic DESs were more soluble than basic DESs. Typically, ChCl: OA isolated lignin successfully, but ChCl: LA isolated lignin less effectively than ChCl: OA; ChCl: Gly, ChCl: EG, and ChCl: Urea. Whereas ChCl: LA and ChCl: OA successfully enhanced the crystallinity of bagasse fibers, 62.3 and 61.7%, respectively. For different lignocellulose biomasses, rice husk and sugar cane bagasse [16], N. Kumar et al. studied the potential of five different deep eutectic solvents for delignification of rice husk and bagasse. The research work investigated that DES composed of ChCl: FA, ChCl: LA, and ChCl: LA was effective. The other two DESs (betaine: lactate and proline: lactate) had showed poor delignification due to weak acidity of DES, but DES with higher acidity (lower pKa) delignified lignin more effectively [102].

Lin et al. processed bamboo waste with various DES molar ratios of ChCl: LA. The findings revealed that increasing the ratio of choline chloride to lactic acid from 2:1 to 1:8 improved the lignin isolation capacity from 42.3 to 80.3% at temperatures of 110 and 130°C, respectively [103, 104]. Similarly, H. Xu, et al. analyzed the effects of DES components' molar ratio using DES of ChCl: LA. The study revealed that delignification decreased with increasing lactic acid content and on the other side, an increase in the monocarboxylic acid hydrogen bonding donor, resulting in enhanced delignification and a decrease in molecular chain length [93]. Whereas, Tan et al. studied the effects of types of DES on delignification of oil palm fruit bunch employing five different acidic deep eutectic solvents; ChCl: LA, ChCl: MA, ChCl: CA, ChCl: FA, and ChCl: AA. The results demonstrated that the presence of hydroxyl groups and short alkyl chains improved lignin extraction. A 1:15 ratio of choline chloride to lactic acid (ChCl: LA) and a 1:2 ratio of choline chloride to formic acid (ChCl: FA) isolated more than 60% of lignin [98, 99].

T. Li et al. also performed experiments to investigate the type of DES and reaction temperature effects on the isolation of lignin from willow. The types of DESs were glycerol, urea, and lactic acid mixed with choline chloride at fixed molar ratio (1:2). Whereas the reaction temperature (90, 100, 110, and 120°C for 6 h) was another parameter that was considered to investigate the treatment effects on lignin yield. The final result reported that the amount of lignin isolated varied with the type of DES, as reportedly high amount of lignin was obtained from DES of ChCl: LA [75]. In addition, it was also investigated that lignin yield varied with reaction temperature and time, which increased with increasing temperature and *vice versa*. Generally, it is investigated that the ChCl: LA pretreatment was found to be much more suitable compared to ChCl: Gly and ChCl: U. It's understood that, in addition to reaction time and temperature, the pretreatment of DES was also affected by the types of deep eutectic solvent components. However, the optimum values of the factor's reaction time, temperature, and molar ratio of DES vary with the type of lignocellulose biomass, as seen from previous research works, which indicates that the type of lignocellulose is also the factor that affects the DES pretreatment. Therefore, optimum pretreatment time, temperature, and molar ratio of one biomass type will not work for the other type of biomass.

5.5 Type of biomass

Deep eutectic solvent (DES) pretreatment has varying effects based on the kind of lignocellulosic biomass feedstock. Alvarez-Vasco et al. investigated the effects of biomass type, such as hardwood and softwood, using DES. The result revealed that the yield of lignin varies with the type of biomass, 78% for hardwood and 58% for softwood [23].

Morán et al. pretreated sugarcane bagasse and beer bagasse using DES of ChCl: GLY. The reported result showed 80 and 15% acid-soluble lignin and Klason lignin isolation, respectively. In addition, it showed increment of saccharification yields with 60, 80, and 100% of dextran, xylan, and arabinoxylan, respectively [96].

For poplar sawdust and herbaceous manzanita, Zhou et al. conducted an experiment employing DES consisting of p-toluene sulfonic acid and choline chloride and exhibited practically full cellulose conversion, with over 90% delignification and xylan elimination [105]. However, thus studies have not determined the reason why the yields of lignin vary with biomass type.

H. Xu et al. conducted an experiment to study the synergistic impacts of various critical factors on DES pretreatment and studied the composition of lignocellulose biomass that hinders DES pretreatment. In the study, 34 different lignocellulosic biomasses were used, including wheat straw, oil palm empty fruit bunches, rice straw, maize straw, and willow jelly. The experiments with biomass feedstocks were investigated using multivariate analysis. The study established that xylan content, lignin content, and dextran content are the most significant components of biomass that influence the effect of DES pretreatment [93]. As seen, the effects of pretreatment vary with the type of biomass, which may be due to the structure of lignin. Thus, it will be a challenge since the structure of lignocellulose biomass varies based on the environment and nature from one area to another, which is also a problem for other pretreatment techniques. For the long-term goal of a renewable economy, genetically engineered biomass types should be planted.

6. Valorization of deep eutectic solvent isolated lignin

Lignin can be upgraded into a number of hydrocarbon fuels and high-value bioproducts (such as aromatic chemicals, macromolecules, advanced materials, or polymers) [106]. Thus, products from lignin can be valorized or processed in different ways depending on the properties of DES lignin. Lignin extracted by deep eutectic solvents has unique properties, such as high purity, low molecular weight, high reactivity, and more preserved β-O-4 bonds, which makes DES lignin interesting for upgrading to aromatic chemical synthesis and synthesis of various polymers. However, these properties may vary with the type of DES. For instance, from the different DES types, carboxylic acid-based DES was the most studied and explored unique feature [4].

Following the properties of extracted lignin, currently there are two categories of lignin valorization such as functionalization and depolymerization [3].

Depolymerization is one of the methods of lignin valorization, which can be carried out by biological [107], thermochemical [108], and electrochemical [109] methods. Lignin's chemical structure may be depolymerized using high and intermediate energetic, catalytic, and biotransformation processes to yield low-value carbon sources for energy generation as well as high-value molecules, such as vanillin and other phenolic compounds.

Functionalization is another way of lignin valorization, described as the activity of improving the reactivity of specific functional groups, such as hydroxyl (OH), methoxyl, carbonyl, carboxyl, or aromatic groups, by hydroxyalkylation, esterification, alkylation, amination, and epoxidation [110]. Following functionalization, lignin can be processed into carbon nanofibers, nanoparticles, resins, polyurethanes, biocomposites, and others [108]. Zdanowicz et al. synthesized a composite thermoplastic starch plasticized by DES of urea: resorcinol: choline chloride (U: R: ChCl, 2:1:1) from potato starch and lignin using twin screw extrusion [111].

Furthermore, because of the strong reactivity *in situ* provided by DES fractionation, DES- isolated lignin might be a promising source for polymer synthesis without additional functionalization [3]. Among other types of DES, carboxylic acid-based DES is the most researched solvent for lignin valorization. Lignin extracted by carboxylic acid-based DES has received a lot of attention because of the distinguishing properties of DES lignin [23]. Lignin isolated from poplar wood by using ChCl: LA has a low molecular weight, an uncondensed structure, and a high purity of lignin [23]. Similarly, Liu et al. isolated low-molecular-weight and high-purity lignin from poplar using microwave-assisted ChCl: OA pretreatment [66].

Another interesting characteristic of lignin is the degree of demethylation, which increases the reactivity of lignin toward chemical alteration. It is shown that lignin extracted from willow via ChCl: LA pretreatment showed more phenolic hydroxyl groups than the other technical lignin, suggesting high reactivity of lignin [80]. Thus, the purity and reactivity of lignin extracted using carboxylic acid-based DES make it appropriate for the production of different polymers [86]. Polyurethane (PU) and phenolic resin are of special importance due to their widespread industrial uses [86, 112]. Kalami et al. produced phenol formaldehyde from agricultural lignin by dilute acid pretreatment and hydrolysis from corn stover [112]. The characteristic properties showed that the produced lignin-phenol resin is similar to commercial phenol resin. Whereas, Xia et al. utilized lignin for the production of biodegradable bioplastic. This work is intended to fill the limitations in mechanical performance and wet stability of bioplastic produced from different biomasses [113]. As a result, the lignocellulosic bioplastic produced has great mechanical strength, excellent water stability, ultraviolet (UV)-light resistance, and increased thermal stability. Furthermore, because it can be easily recycled or biodegraded in the natural environment, lignocellulosic bioplastic has a lesser environmental effect.

7. Conclusions

This book chapter discussed recent advances in lignin fractionation and valorization using deep eutectic solvents (DESs) as a more environmentally friendly alternative to fossil-based synthesis. DESs are promising for lignin extraction due to their chemical and thermal stability, low volatility, and simplicity of synthesis. However, more theoretical and technological studies are needed to better understand the molecular interactions between DESs and lignin, to develop effective DESs recycling methods, and to create more economically feasible lignin biorefinery systems. Tunable DES systems with specific lignin properties for cost-effective valorization are expected in the near future. More research is needed to determine the most effective usage of DES, selection criteria, and operating conditions.

Acknowledgements

The authors acknowledge Helena Recic, Publishing Process Manager in INTECHOPEN for her kindness and endless support. The authors also acknowledge the academic editors and INTECHOPEN for accepting funding wavier.

Conflict of interest

The authors declare no conflicts of interest.

Author details

Tsegaye Abera Mekuriaw* and Belachew Zegale Tizazu
Addis Ababa Science and Technology University, Addis Ababa, Ethiopia

*Address all correspondence to: tsegaye.abera@aastu.edu.et

References

[1] Ullah M, Liu P, Xie S, Sun S. Recent advancements and challenges in lignin valorization: Green routes towards sustainable bioproducts. Molecules. 2022;**27**:6055

[2] Nadányi R, Ház A, Lisý A, Jablonský M, Šurina I, Majová V, et al. Lignin modifications, applications, and possible market prices. Energies. 2022; **15**(18):1-16

[3] Zhou M, Fakayode OA, Ahmed Yagoub AEG, Ji Q, Zhou C. Lignin fractionation from lignocellulosic biomass using deep eutectic solvents and its valorization. Renewable and Sustainable Energy Reviews. Pergamon. 2022;**156**:111986

[4] Chen Z, Ragauskas A, Wan C. Lignin extraction and upgrading using deep eutectic solvents. Industrial Crops and Products. [Internet]. 2020;**147** (February):112241. DOI: 10.1016/j.indcrop.2020.112241

[5] Poveda-Giraldo JA, Solarte-Toro JC, Cardona Alzate CA. The potential use of lignin as a platform product in biorefineries: A review. [Internet]. Renewable and Sustainable Energy Reviews. 2021;**138**:110688. DOI: 10.1016/j.rser.2020.110688

[6] Hong S, Shen XJ, Pang B, Xue Z, Cao XF, Wen JL, et al. In-depth interpretation of the structural changes of lignin and formation of diketones during acidic deep eutectic solvent pretreatment. Green Chemistry. 2020; **22**(6):1851-1858

[7] El Achkar T, Greige-Gerges H, Fourmentin S. Basics and properties of deep eutectic solvents: A review. [Internet]. Environmental Chemistry Letters. 2021;**19**:3397-3408. DOI: 10.1007/s10311-021-01225-8

[8] Smith EL, Abbott AP, Ryder KS. Deep eutectic solvents (DESs) and their applications. Chemical Reviews. 2014; **114**(21):11060-11082

[9] Zhang M, Zhang X, Liu Y, Wu K, Zhu Y, Lu H, et al. Insights into the relationships between physicochemical properties, solvent performance, and applications of deep eutectic solvents. Environmental Science and Pollution Research. 2021;**28**(27):35537-35563

[10] Hong S, Shen XJ, Xue Z, Sun Z, Yuan TQ. Structure-function relationships of deep eutectic solvents for lignin extraction and chemical transformation. Green Chemistry. 2020; **22**(21):7219-7232

[11] Van ODJGP, Van SJ, Esteves ACC, Tuinier R, Vis M. Promising technological and industrial applications of deep EutecticEutectic systems. Physical Chemistry Chemical Physics [Internet]. 2021;**22**:2181-2187. DOI: 10.1039/C9CP06762K

[12] Abbott AP, Barron JC, Ryder KS, Wilson D. Eutectic-based ionic liquids with metal-containing anions and cations. Chemistry : A European Journal. 2007;**13**(22):6495-6501

[13] Meraj A, Singh SP, Jawaid M, Nasef MM, Alomar TS, AlMasoud N. A review on eco-friendly isolation of lignin by natural deep eutectic solvents from agricultural wastes. [Internet]. Journal of Polymers and the Environment. 2023;**31** (8). DOI: 10.1007/s10924-023-02817-x

[14] Florindo C, Branco LC, Marrucho IM. Quest for green-solvent design: From hydrophilic to hydrophobic (deep) eutectic solvents. ChemSusChem. 2019;**12**(8):1549-1559

[15] Abbott AP, Capper G, Davies DL, Rasheed RK. Ionic liquid analogues formed from hydrated metal salts. Chemistry : A European Journal. 2004; **10**(15):3769-3774

[16] Cao J, Su E. Hydrophobic deep eutectic solvents: The new generation of green solvents for diversified and colorful applications in green chemistry. Journal of Cleaner Production [Internet]. 2021;**314**(December 2020):127965. DOI: 10.1016/j.jclepro.2021.127965

[17] Gurkan B, Squire H, Pentzer E. Metal-free deep eutectic solvents: Preparation, physical properties, and significance. Journal of Physical Chemistry Letters. 2019;**10**(24): 7956-7964

[18] Santana APR, Mora-Vargas JA, Guimarães TGS, Amaral CDB, Oliveira A, Gonzalez MH. Sustainable synthesis of natural deep eutectic solvents (NADES) by different methods. Journal of Molecular Liquids. 2019;**293**:17-20

[19] Abranches DO, Martins MAR, Silva LP, Schaeffer N, Pinho SP, Coutinho JAP. Phenolic hydrogen bond donors in the formation of non-ionic deep eutectic solvents: The quest for type v des. Chemical Communications. 2019;**55**(69):10253-10256

[20] Li P, Zhang Z, Zhang X, Li K, Jin Y, Wu W. DES: Their effect on lignin and recycling performance. RSC Advances. 2023;**13**(5):3241-3254

[21] Kohli K, Katuwal S, Biswas A, Sharma BK. Effective delignification of lignocellulosic biomass by microwave assisted deep eutectic solvents. Bioresource Technology. 2020;**303**: 122897

[22] Liu Q, Zhao X, Yu D, Yu H, Zhang Y, Xue Z, et al. Novel deep eutectic solvents with different functional groups towards highly efficient dissolution of lignin. Green Chemistry. 2019;**21**(19):5291-5297

[23] Alvarez-Vasco C, Ma R, Quintero M, Guo M, Geleynse S, Ramasamy KK, et al. Unique low-molecular-weight lignin with high purity extracted from wood by deep eutectic solvents (DES): A source of lignin for valorization. Green Chemistry. 2016;**18**(19):5133-5141

[24] Liu Q, Mou H, Chen W, Zhao X, Yu H, Xue Z, et al. Highly efficient dissolution of lignin by eutectic molecular liquids. Industrial and Engineering Chemistry Research. 2019; **58**(51):23438-23444

[25] Salehi HS, Ramdin M, Moultos OA, Vlugt TJH. Computing solubility parameters of deep eutectic solvents from molecular dynamics simulations. Fluid Phase Equilibria [Internet]. 2019; **497**:10-18. DOI: 10.1016/j.fluid.2019.05.022

[26] Zhang Q, De Oliveira VK, Royer S, Jérôme F. Deep eutectic solvents: Syntheses, properties and applications. Chemical Society Reviews. 2012;**41**(21): 7108-7146

[27] Hou YC, Yao CF, Wu WZ. Deep eutectic solvents: Green solvents for separation applications. Wuli Huaxue Xuebao/Acta Physico-Chimica Sinica. 2018;**34**(8):873-885

[28] Liu P, Hao JW, Mo LP, Zhang ZH. Recent advances in the application of deep eutectic solvents as sustainable media as well as catalysts in organic reactions. RSC Advances [Internet]. 2015;**5**(60):48675-48704. DOI: 10.1039/C5RA05746A

[29] Abbott AP, Capper G, Davies DL, Rasheed RK, Tambyrajah V. Novel

solvent properties of choline chloride/ urea mixtures. Chemical Communications. 2003;**1**:70-71

[30] Makoś P, Przyjazny A, Boczkaj G. Hydrophobic deep eutectic solvents as "green" extraction media for polycyclic aromatic hydrocarbons in aqueous samples. Journal of Chromatography. A. 2018;**1570**:28-37

[31] Satlewal A, Agrawal R, Bhagia S, Sangoro J, Ragauskas AJ. Natural deep eutectic solvents for lignocellulosic biomass pretreatment: Recent developments, challenges and novel opportunities. Biotechnology Advances. 2018;**36**(8):2032-2050. DOI: 10.1016/j.biotechadv.2018.08.009

[32] Abbott AP, Boothby D, Capper G, Davies DL, Rasheed RK. Deep eutectic solvents formed between choline chloride and carboxylic acids: Versatile alternatives to ionic liquids. Journal of the American Chemical Society. 2004; **126**(29):9142-9147

[33] Cui Y, Li C, Yin J, Li S, Jia Y, Bao M. Design, synthesis and properties of acidic deep eutectic solvents based on choline chloride. Journal of Molecular Liquids. 2017;**236**:338-343

[34] Alomar MK, Hayyan M, Alsaadi MA, Akib S, Hayyan A, Hashim MA. Glycerol-based deep eutectic solvents: Physical properties. Journal of Molecular Liquids. 2016;**215**:98-103

[35] Hayyan A, Mjalli FS, Alnashef IM, Al-Wahaibi YM, Al-Wahaibi T, Hashim MA. Glucose-based deep eutectic solvents: Physical properties. Journal of Molecular Liquids [Internet]. 2013;**178**:137-141. DOI: 10.1016/j.molliq.2012.11.025

[36] García G, Aparicio S, Ullah R, Atilhan M. Deep eutectic solvents: Physicochemical properties and gas separation applications. Energy and Fuels. 2015;**29**(4):2616-2644

[37] Shah D, Mjalli FS. Effect of water on the thermo-physical properties of reline: An experimental and molecular simulation based approach. Physical Chemistry Chemical Physics. 2014;**16** (43):23900-23907

[38] Bara JE, Camper DE, Gin DL, Noble RD. Room-temperature ionic liquids and composite materials: Platform technologies for CO_2 capture. Accounts of Chemical Research. 2009;**43**(1):152-159

[39] Yadav A, Trivedi S, Rai R, Pandey S. Densities and dynamic viscosities of (choline chloride+glycerol) deep eutectic solvent and its aqueous mixtures in the temperature range (283.15-363.15)K. Fluid Phase Equilibria [Internet]. 2014; **367**:135-142. DOI: 10.1016/j.fluid.2014.01.028

[40] Guo W, Hou Y, Ren S, Tian S, Wu W. Formation of deep eutectic solvents by phenols and choline chloride and their physical properties. Journal of Chemical & Engineering Data. 2013;**58** (4):866-872

[41] Mjalli FS, Jabbar NMA. Acoustic investigation of choline chloride based ionic liquids analogs. Fluid Phase Equilibria [Internet]. 2014;**381**:71-76. DOI: 10.1016/j.fluid.2014.08.017

[42] Florindo C, Oliveira FS, Rebelo LPN, Fernandes AM, Marrucho IM. Insights into the synthesis and properties of deep eutectic solvents based on cholinium chloride and carboxylic acids. ACS Sustainable Chemistry & Engineering. 2014;**2**(10):2416-2425

[43] Verma R, Mohan M, Goud VV, Banerjee T. Operational strategies and

comprehensive evaluation of menthol based deep eutectic solvent for the extraction of lower alcohols from aqueous media. ACS Sustainable Chemistry & Engineering. 2018;**6**(12): 16920-16932

[44] Dai Y, Witkamp GJ, Verpoorte R, Choi YH. Tailoring properties of natural deep eutectic solvents with water to facilitate their applications. Food Chemistry [Internet]. 2015;**187**:14-19. DOI: 10.1016/j.foodchem.2015.03.123

[45] Widsten P, Tamminen T, Liitiä T. Natural sunscreens based on nanoparticles of modified Kraft lignin (CatLignin). ACS Omega. 2020;**5**(22): 13438-13446

[46] Nunes RJ, Saramago B, M. Marrucho I. Surface tension of dl-menthol: Octanoic acid eutectic mixtures. Journal of Chemical & Engineering Data. 2019;**64**(11): 4915-4923

[47] D'Agostino C, Harris RC, Abbott AP, Gladden LF, Mantle MD. Molecular motion and ion diffusion in choline chloride based deep eutectic solvents studied by 1H pulsed field gradient NMR spectroscopy. Physical Chemistry Chemical Physics. 2011;**13**(48): 21383-21391

[48] Mjalli FS, Naser J, Jibril B, Alizadeh V, Gano Z. Tetrabutylammonium chloride based ionic liquid analogues and their physical properties. Journal of Chemical & Engineering Data. 2014;**59**(7):2242-2251

[49] Jibril B, Mjalli F, Naser J, Gano Z. New tetrapropylammonium bromide-based deep eutectic solvents: Synthesis and characterizations. Journal of Molecular Liquids [Internet]. 2014;**199**: 462-469. DOI: 10.1016/j.molliq.2014.08.004

[50] Reichardt C. Solvatochromic dyes as solvent polarity indicators. Chemical Reviews. 2002;**94**(8):2319-2358

[51] Kamlet MJ, Luis Abboud J, Taft RW. The solvatochromic comparison method. 6. The .pi.* scale of solvent polarities. Journal of the American Chemical Society. 2002;**99**(18):6027-6038

[52] Kamlet JM, Taft WR. The solvatochromic comparison method. I. The .Beta.-scale of solvent hydrogen-bond acceptor (HBA) basicities. Journal of the American Chemical Society. 2002; **98**(2):377-383

[53] Abbott AP, Harris RC, Ryder KS, D'Agostino C, Gladden LF, Mantle MD. Glycerol eutectics as sustainable solvent systems. Green Chemistry. 2011;**13**(1): 82-90

[54] Valvi A, Dutta J, Tiwari S. Temperature-dependent empirical parameters for polarity in choline chloride based deep eutectic solvents. The Journal of Physical Chemistry. B. 2017;**121**(50):11356-11366

[55] Azizi N, Dezfooli S, Hashemi MM. A sustainable approach to the Ugi reaction in deep eutectic solvent. Comptes Rendus Chimie. 2013;**16**(12):1098-1102

[56] Mahto A, Mondal D, Polisetti V, Bhatt J, Nidhi MR, Prasad K, et al. Sustainable water reclamation from different feed streams by forward osmosis process using deep eutectic solvents as reusable draw solution. Industrial and Engineering Chemistry Research. 2017;**56**(49):14623-14632

[57] Farooq MQ, Abbasi NM, Anderson JL. Deep eutectic solvents in separations: Methods of preparation, polarity, and applications in extractions and capillary electrochromatography. Journal of Chromatography. A

[Internet]. 2020;**1633**:461613. DOI: 10.1016/j.chroma.2020.461613

[58] Gomez FJV, Espino M, Fernández MA, Silva MF. A greener approach to prepare natural deep eutectic solvents. ChemistrySelect. 2018; **3**(22):6122-6125

[59] Rodriguez Rodriguez N, van den Bruinhorst A, Kollau LJBM, Kroon MC, Binnemans K, et al. Degradation of deep-eutectic solvents based on choline chloride and carboxylic acids. ACS Sustainable Chemistry & Engineering. 2019;**7**(13):11521-11528

[60] Sekharan TR, Chandira RM, Tamilvanan S, Rajesh SC, Venkateswarlu BS. Deep eutectic solvents as an alternate to other harmful solvents. Biointerface Research in Applied Chemistry. 2022;**12**(1):847-860

[61] Zhang M, Tian R, Han H, Wu K, Wang B, Liu Y, et al. Preparation strategy and stability of deep eutectic solvents: A case study based on choline chloride-carboxylic acid. Journal of Cleaner Production. 2022;**345**:131028

[62] Espino M, de los Ángeles Fernández M, FJV G, Silva MF. Natural designer solvents for greening analytical chemistry. TrAC Trends in Analytical Chemistry [Internet]. 2016;**76**:126-136. DOI: 10.1016/j.trac.2015.11.006

[63] Gutiérrez MC, Ferrer ML, Mateo CR, Del MF. Freeze-drying of aqueous solutions of deep eutectic solvents: A suitable approach to deep eutectic suspensions of self-assembled structures. Langmuir. 2009;**25**(10): 5509-5515

[64] Hsieh YH, Li Y, Pan Z, Chen Z, Lu J, Yuan J, et al. Ultrasonication-assisted synthesis of alcohol-based deep eutectic solvents for extraction of active compounds from ginger. Ultrasonics Sonochemistry [Internet]. 2020;**63**: 104915. DOI: 10.1016/j.ultsonch.2019. 104915

[65] Dai Y, van Spronsen J, Witkamp GJ, Verpoorte R, Choi YH. Natural deep eutectic solvents as new potential media for green technology. Analytica Chimica Acta [Internet]. 2013;**766**:61-68. DOI: 10.1016/j.aca.2012.12.019

[66] Liu YY, Chen W, Xia Q, Guo B, Wang Q, Liu S, et al. Efficient cleavage of lignin–carbohydrate complexes and ultrafast extraction of lignin oligomers from Wood biomass by microwave-assisted treatment with deep eutectic solvent. ChemSusChem. 2017;**10**(8): 1692-1700

[67] Kumar AK, Parikh BS, Pravakar M. Natural deep eutectic solvent mediated pretreatment of rice straw: Bioanalytical characterization of lignin extract and enzymatic hydrolysis of pretreated biomass residue. Environmental Science and Pollution Research [Internet]. 2016; **23**(10):9265-9275. DOI: 10.1007/ s11356-015-4780-4

[68] Mankar AR, Pandey A, Modak A, Pant KK. Pretreatment of lignocellulosic biomass: A review on recent advances. Bioresource Technology [Internet]. 2021;**334**(March):125235. DOI: 10.1016/j. biortech.2021.125235

[69] Borrero-l AM, Valencia C, Franco JM. Lignocellulosic materials for the production of biofuels, biochemicals and biomaterials and applications of. Polymers (Basel). 2022;**14**(5):881. DOI: 10.3390/polym14050881

[70] Kumar B, Bhardwaj N, Agrawal K, Chaturvedi V, Verma P. Current perspective on pretreatment technologies using lignocellulosic

biomass: An emerging biorefinery concept. Fuel Processing Technology. 2020;**199**:106244

[71] Shah AA, Seehar TH, Sharma K, Toor SS. Biomass pretreatment technologies. Hydrocarb Biorefinery Sustain Process Biomass Hydrocarb Biofuels. 2022:203-228

[72] Baruah J, Nath BK, Sharma R, Kumar S, Deka RC, Baruah DC, et al. Recent trends in the pretreatment of lignocellulosic biomass for value-added products. Frontiers in Energy Research. 2018;**6**(Dec):1-19

[73] Xu X, Wang K, Zhou Y, Lai C, Zhang D, Xia C, et al. Comparison of organosolv pretreatment of masson pine with different solvents in promoting delignification and enzymatic hydrolysis efficiency. Fuel. 2023;**338**: 127361

[74] Zhao ZM, Meng X, Scheidemantle B, Pu Y, Liu ZH, Li BZ, et al. Cosolvent enhanced lignocellulosic fractionation tailoring lignin chemistry and enhancing lignin bioconversion. Bioresource Technology. 2022;**347**:126367

[75] Tomé LIN, Baião V, da Silva W, Brett CMA. Deep eutectic solvents for the production and application of new materials. Applied Materials Today. 2018;**10**:30-50

[76] Malaeke H, Housaindokht MR, Monhemi H, Izadyar M. Deep eutectic solvent as an efficient molecular liquid for lignin solubilization and wood delignification. Journal of Molecular Liquids. 2018;**263**:193-199

[77] Zhang CW, Xia SQ, Ma PS. Facile pretreatment of lignocellulosic biomass using deep eutectic solvents. Bioresource Technology [Internet]. 2016;**219** (July):1-5. DOI: 10.1016/j.biortech.2016.07.026

[78] Chen Z, Wan C. Ultrafast fractionation of lignocellulosic biomass by microwave-assisted deep eutectic solvent pretreatment. Bioresource Technology. 2018;**250**:532-537

[79] Shen XJ, Wen JL, Mei QQ, Chen X, Sun D, Yuan TQ, et al. Facile fractionation of lignocelluloses by biomass-derived deep eutectic solvent (DES) pretreatment for cellulose enzymatic hydrolysis and lignin valorization. Green Chemistry. 2019; **21**(2):275-283

[80] Zhao. Organosolv fractionating pre-treatment of lignocellulosic biomass for efficient enzymatic saccharification: Chemistry, kinetics, and substrate structures. Biofuels, Bioproducts and Biorefining. 2017;**8**(6):24

[81] Chen Z, Jacoby WA, Wan C. Ternary deep eutectic solvents for effective biomass deconstruction at high solids and low enzyme loadings. Bioresource Technology. 2019;**279**: 281-286

[82] Xu F, Sun J, Wehrs M, Ho Kim K, Rau SS, Chan AM, et al. Biocompatible choline-based deep eutectic solvents enable one-pot production of cellulosic ethanol. ACS Sustainable Chemistry & Engineering. 2018;**6**(7):8914-8919

[83] Procentese A, Johnson E, Orr V, Garruto Campanile A, Wood JA, Marzocchella A, et al. Deep eutectic solvent pretreatment and subsequent saccharification of corncob. Bioresource Technology. 2015;**192**:31-36

[84] Chen Z, Bai X, Lusi A, Wan C. High-solid lignocellulose processing enabled by natural deep eutectic solvent for lignin extraction and industrially

relevant production of renewable chemicals. ACS Sustainable Chemistry & Engineering. 2018;**6**(9):12205-12216

[85] Zhao Z, Chen X, Ali MF, Abdeltawab AA, Yakout SM, Yu G. Pretreatment of wheat straw using basic ethanolamine-based deep eutectic solvents for improving enzymatic hydrolysis. Bioresource Technology. 2018;**263**:325-333

[86] Pang B, Yang S, Fang W, Yuan TQ, Argyropoulos DS, Sun RC. Structure-property relationships for technical lignins for the production of lignin-phenol-formaldehyde resins. Industrial Crops and Products. 2017;**108**:316-326

[87] Kim KH, Dutta T, Sun J, Simmons B, Singh S. Biomass pretreatment using deep eutectic solvents from lignin derived phenols. Green Chemistry. 2018; **20**(4):809-815

[88] Liu Y, Zheng J, Xiao J, He X, Zhang K, Yuan S, et al. Enhanced enzymatic hydrolysis and lignin extraction of wheat straw by Triethylbenzyl ammonium chloride/ lactic acid-based deep eutectic solvent Pretreatment. ACS Omega. 2019;**4**(22): 19829-19839

[89] Tan YT, Ngoh GC, Chua ASM. Evaluation of fractionation and delignification efficiencies of deep eutectic solvents on oil palm empty fruit bunch. Industrial Crops and Products. 2018;**123**:271-277

[90] Hou XD, Lin KP, Li AL, Yang LM, Fu MH. Effect of constituents molar ratios of deep eutectic solvents on rice straw fractionation efficiency and the micro-mechanism investigation. Industrial Crops and Products. 2018;**120**:322-329

[91] Guo Z, Zhang Q, You T, Zhang X, Xu F, Wu Y. Short-time deep eutectic solvent pretreatment for enhanced enzymatic saccharification and lignin valorization. Green Chemistry. 2019; **21**(11):3099-3108

[92] Xing W, Xu G, Dong J, Han R, Ni Y. Novel dihydrogen-bonding deep eutectic solvents: Pretreatment of rice straw for butanol fermentation featuring enzyme recycling and high solvent yield. Chemical Engineering Journal. 2018;**333**: 712-720

[93] Xu H, Kong Y, Peng J, Song X, Liu Y, Su Z, et al. Comprehensive analysis of important parameters of choline chloride-based deep eutectic solvent pretreatment of lignocellulosic biomass. Bioresource Technology [Internet]. 2021;**319**:124209. DOI: 10.1016/j. biortech.2020.124209

[94] Procentese A, Raganati F, Olivieri G, Russo ME, Rehmann L, Marzocchella A. Deep eutectic solvents pretreatment of agro-industrial food waste. Biotechnology for Biofuels and Bioproducts [Internet]. 2018;**11**(1):1-12. DOI: 10.1186/s13068-018-1034-y

[95] Ma H, Fu P, Zhao J, Lin X, Wu W, Yu Z, et al. Pretreatment of wheat straw lignocelluloses by deep eutectic solvent for lignin extraction. Molecules. 2022;**27** (22):7955

[96] Morán-aguilar MG, Costa-trigo I, Ramírez-pérez AM, de Blas E, Calderón-santoyo M, Aguilar-uscanga MG, et al. Development of sustainable biorefinery processes applying deep eutectic solvents to Agrofood wastes. Energies. 2022;**15**(11):1-12

[97] Lou R, Ma R, Lin KT, Ahamed A, Zhang X. Facile extraction of wheat straw by deep eutectic solvent (DES) to produce lignin nanoparticles. ACS Sustainable Chemistry & Engineering. 2019;7(12):10248-10256

[98] Tan YT, Ngoh GC, Chua ASM. Effect of functional groups in acid constituent of deep eutectic solvent for extraction of reactive lignin. Bioresource Technology [Internet]. 2019;**281**:359-366. DOI: 10.1016/j.biortech.2019.02.010

[99] Li C, Huang C, Zhao Y, Zheng C, Su H, Zhang L, et al. Effect of choline-based deep eutectic solvent pretreatment on the structure of cellulose and lignin in bagasse. PRO. 2021;**9**(2):1-14

[100] Li T, Lyu G, Liu Y, Lou R, Lucia LA, Yang G, et al. Deep eutectic solvents (DESs) for the isolation of willow lignin (Salix matsudana cv. Zhuliu). International Journal of Molecular Sciences. 2017;**18**(11):2266

[101] Okur M, Eslek Koyuncu DD. Investigation of pretreatment parameters in the delignification of paddy husks with deep eutectic solvents. Biomass and Bioenergy [Internet]. 2020; **142**(July):105811. DOI: 10.1016/j.biombioe.2020.105811

[102] Kumar N, Gautam R, Stallings JD, Coty GG, Lynam JG. Secondary agriculture residues Pretreatment using deep eutectic solvents. Waste and Biomass Valorization [Internet]. 2021;**12**(5):2259-2269. DOI: 10.1007/s12649-020-01176-1

[103] Lin W, Xing S, Jin Y, Lu X, Huang C, Yong Q. Insight into understanding the performance of deep eutectic solvent pretreatment on improving enzymatic digestibility of bamboo residues. Bioresource Technology [Internet]. 2020;**306**(March):123163. DOI: 10.1016/j.biortech.2020.123163

[104] Xu H, Kong Y, Peng J, Wang W, Li B. Mechanism of deep eutectic solvent delignification: Insights from molecular dynamics simulations. ACS Sustainable Chemistry & Engineering. 2021;**9**(20):7101-7111

[105] Zhou X, Huang T, Liu J, Gao H, Bian H, Wang R, et al. Recyclable deep eutectic solvent coupling sodium hydroxide post-treatment for boosting woody/herbaceous biomass conversion at mild condition. Bioresource Technology [Internet]. 2021;**320**(PA):124327. DOI: 10.1016/j.biortech.2020.124327

[106] Schutyser W, Renders T, Van Den Bosch S, Koelewijn SF, Beckham GT, Sels BF. Chemicals from lignin: An interplay of lignocellulose fractionation, depolymerisation, and upgrading. Chemical Society Reviews [Internet]. 2018;**47**(3):852-908. DOI: 10.1039/C7CS00566K

[107] Xu R, Zhang K, Liu P, Han H, Zhao S, Kakade A, et al. Lignin depolymerization and utilization by bacteria. Bioresource Technology. [Internet]. 2018;**269**:557-566. DOI: 10.1016/j.biortech.2018.08.118

[108] Zevallos Torres LA, Lorenci Woiciechowski A, de Andrade Tanobe VO, Karp SG, Guimarães Lorenci LC, Faulds C, et al. Lignin as a potential source of high-added value compounds: A review. Journal of Cleaner Production. 2020;**263**:121499

[109] Fache M, Boutevin B, Caillol S. Vanillin production from lignin and its use as a renewable chemical. ACS Sustainable Chemistry & Engineering. 2016;**4**(1):35-46

[110] Tran MH, Lee EY. Green preparation of bioplastics based on degradation and chemical modification of lignin residue. Journal of Wood Chemistry and Technology [Internet]. 2018;**38**(6):460-478. DOI: 10.1080/02773813.2018.1533978

[111] Zdanowicz M, Sałasińska K, Lewandowski K, Skórczewska K. Thermoplastic starch/ternary deep eutectic solvent/lignin materials: Study of physicochemical properties and fire behavior. ACS Sustainable Chemistry & Engineering. 2022;**10**(14):4579-4587

[112] Kalami S, Arefmanesh M, Master E, Nejad M. Replacing 100% of phenol in phenolic adhesive formulations with lignin. Journal of Applied Polymer Science. 2017;**134**(30):2-10

[113] Xia Q, Chen C, Yao YY, Li J, He S, Zhou Y, et al. A strong, biodegradable and recyclable lignocellulosic bioplastic. Nature Sustainability [Internet]. 2021;**4** (7):627-635. DOI: 10.1038/s41893-021-00702-w

Section 2

Advanced Applications

Chapter 3

Advanced Lignin Valorization for Biorefinery Application

Filemon Jalu Nusantara Putra, Prihardi Kahar, Chiaki Ogino and Akihiko Kondo

Abstract

Despite concerns about diminishing fossil fuels and the imperative for renewable alternatives, lignin, Earth's most abundant aromatic biopolymer, remains largely underutilized. This chapter explores the immense potential of lignin biorefinery to address energy demands, promote economic growth, and adhere to sustainable development principles. However, intricate structure, harsh odor, and toxicity hinder its valorization. To address these issues, examining emerging biochemical strategies, including thermochemical and enzymatic depolymerization and physical techniques, have emerged as promising avenues for converting lignin into valuable biofuels and chemical compounds. By highlighting innovative approaches and technologies, it emphasizes the pivotal role of lignin in driving biorefineries toward low-emission processes, yielding a diverse spectrum of bio-products. This chapter aims to contribute to the ongoing discourse on sustainable and eco-friendly biorefinery practices of lignin valorization.

Keywords: lignin, depolymerization, biorefinery, sustainable valorization, biofuels, bioproducts

1. Introduction

In recent years, growing concerns about the depletion of fossil fuels have ignited a quest for sustainable and renewable alternatives to meet the escalating global energy demands [1–3]. Amidst this backdrop, lignin, the most abundant aromatic biopolymer on Earth, emerges as a promising solution, representing an impressive reservoir of approximately 300 billion metric tons [4, 5]. Leveraging lignin for producing biofuels and diverse chemicals through biorefinery processes holds the potential to not only alleviate the strain on diminishing fossil fuel resources but also propel economic growth while adhering to principles of sustainable development. However, the realization of lignin's full potential faces formidable challenges. A significant portion of lignin is relegated to combustion, a practice that fails to harness its intrinsic value [6]. This predicament is primarily attributed to lignin's intricate and rigid structures, pungent odor, and toxicity to certain organisms, rendering sustainable valorization a complex endeavor [7]. As shown in **Figure 1**, biochemical strategies, such as thermochemical coupled with physical techniques and microbial enzymatic

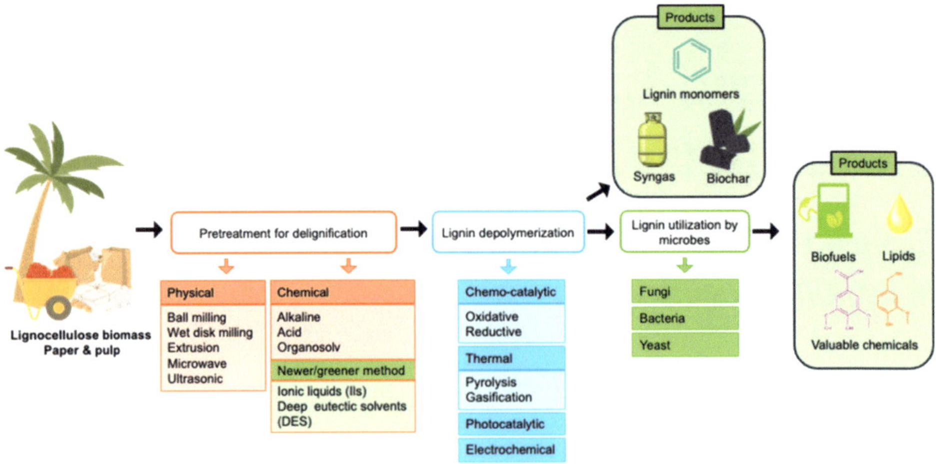

Figure 1.
Overview of lignin valorization for biorefinery application.

depolymerization, have emerged as promising avenues for converting lignin into value-added compounds [8].

This book chapter endeavors to shed light on the existing challenges and breakthroughs in the sustainable valorization of lignin for biorefinery applications. By delving into innovative approaches and technologies, the chapter aims to underscore the pivotal role of lignin in steering future biorefineries toward low-emission processes, yielding a spectrum of valuable biofuels and chemical compounds. This chapter aspires to contribute to the ongoing discourse on lignin utilization for sustainable and eco-friendly biorefinery practices by comprehensively exploring the current landscape.

2. Lignin as a valorization material

2.1 Structure and properties

The intricate structure of lignin as a natural macromolecule comprises methoxy groups, phenolic hydroxyl groups, and a few terminal aldehyde groups. These functional groups are attached to a complex network of phenylpropanoid units, the fundamental building blocks of lignin [9]. The three main types of phenylpropanoids in lignin are p-hydroxyphenyl (H), guaiacyl (G), and syringyl (S), each distinguished by the degree of methoxylation on their aromatic rings (**Figure 2**) [10]. The aromatic rings of phenylpropanoid units, abundant in lignin, impart antioxidant activity [11]. These monolignols undergo random coupling through various linkages, and this coupling process results in a diverse array of C-O-C and C-C linkages. The linkages include α-O-4, β-O-4, 4-O-5 (C-O-C) and β-β, β-1, β-5, 5-5 (C-C) [12, 13]. Primarily, β-O-4 forms the three-dimensional structure of lignin, accounting for up to 50% of all bonds [14, 15]. Unlike cellulose and hemicellulose, where the biopolymer is readily decomposed into sugars [16], the highly polymerized lignin exhibits a rigid structure that resists chemical pretreatment [17]. Cleaving the β-O-4 linkage has been a crucial target for lignin depolymerization strategies. The source of biomass strongly

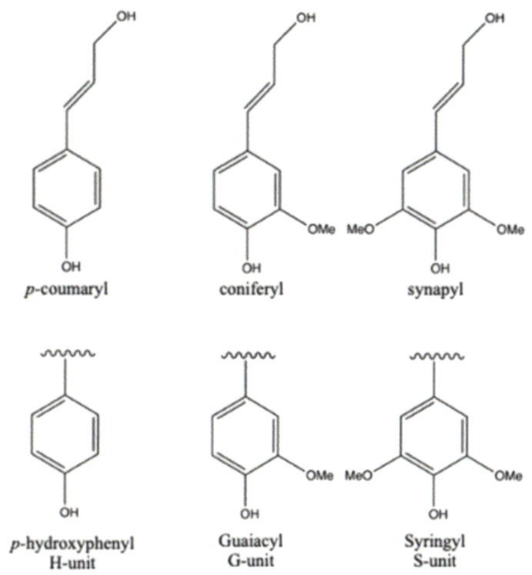

Figure 2.
Relative structures of group units of lignin.

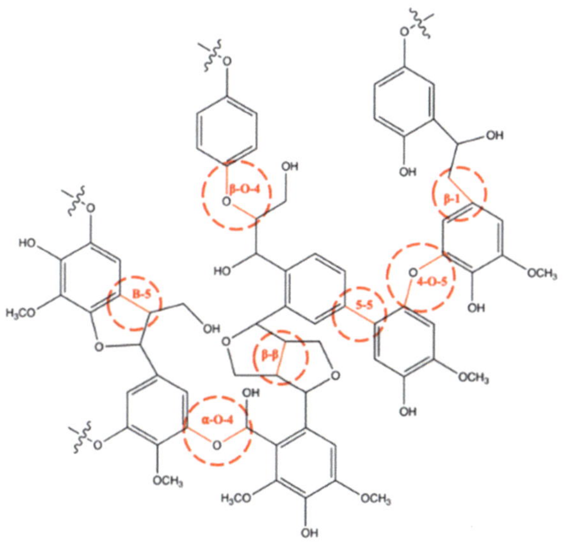

Figure 3.
Lignin polymer structure and its main linkages.

influences the structure and composition of lignin. For instance, softwood lignin contains a high proportion of G units (80–90%), hardwood lignin contains a mixture of G (25–50%) and S (50–70%) units, and grass lignin comprises a mix of G (25–50%), H (10–25%), and S (25–50%) aromatic units [14, 18]. Consistently, S/G-rich lignin is preferred for maximizing monomer yields in biorefining applications due to its ability to reduce the formation of inactive C-C bonds, leading to undesired condensed lignin during lignin depolymerization. High G and low S content in lignin could lead to a surge in β-O-4 linkages, creating readily recombining monomers during fractionation [19, 20]. Understanding lignin's structure and identifying its native linkages are essential for its efficient valorization into renewable fuels, materials, or chemicals (**Figure 3**).

2.2 Lignin characterization

To effectively utilize the lignin from the abundant type of lignocellulosic biomass (LB), it is essential to understand its intricate structure and properties. The various pretreatment methods employed for lignin extraction can drastically alter its structure, composition, and properties. Several factors, including the origin of the lignocellulosic biomass, the pretreatment process, and the specific isolation and purification techniques used, influence the extent of these modifications. Therefore, to effectively valorize the extracted lignin into valuable aromatic compounds, it is crucial to understand the precise changes induced in its physicochemical properties through meticulous characterization techniques.

Various characterization techniques play a vital role in unveiling lignin's secrets. Two prominent techniques are gel permeation chromatography (GPC) and Fourier transform infrared spectroscopy (FTIR). GPC measures molecular weight distribution, revealing how pretreatment methods like THF-based extraction affect lignin's size [21]. FTIR provides information on functional groups present in the structure. The presence of specific bands, such as those corresponding to aromatic O−H and methoxy groups, confirms the integrity of the lignin structure [22]. Determining thermal stability is crucial for various applications. Thermogravimetric analysis (TGA) and differential scanning calorimetry (DSC) offer valuable insights. TGA reveals the weight loss pattern as temperature increases, indicating the presence of different chemical bonds and their decomposition temperatures [23]. DSC helps determine glass transition temperature (Tg) and heat of pyrolysis. Higher Tg values suggest better thermal stability, with lignin boasting higher Tg compared to synthetic polymers due to its complex structure [24]. NMR spectroscopy shines in unraveling the detailed structure of lignin. ^{1}H NMR provides information about different functional groups based on proton signals. ^{13}C NMR offers better resolution and facilitates the quantification of specific linkages like β-O-4 and β-β bonds [25]. ^{31}P NMR allows rapid quantification of hydroxyl groups using phosphorylation. 2D HSQC NMR or 2D ^{1}H-^{13}C heteronuclear single quantum coherence NMR provides high resolution, enabling the identification of different monolignol units, their ratios, and the relative abundance of interunit linkages [26]. By combining the information gleaned from these diverse characterization techniques, researchers can profoundly understand lignin's structure and properties. This knowledge paves the way for its effective utilization in various applications, ranging from biofuels and bioplastics to sustainable materials and pharmaceuticals.

3. Pretreatment for lignocellulosic biomass delignification

3.1 The power of physical pretreatment

Physical pretreatment is the essential first step in unlocking the potential of lignocellulosic biomass for biofuel production. By reducing particle size, increasing surface area, and altering crystallinity, it paves the way for efficient subsequent processes like enzymatic hydrolysis and fermentation [27]. While energy consumption remains a challenge, with some methods requiring significant power, the advantages of physical pretreatment are undeniable. Its eco-friendly nature, minimal toxic byproduct generation, and effectiveness on a wide variety of feedstocks make it a crucial tool in the biofuel toolbox [28]. Ongoing research in optimizing energy efficiency, like the use of wet disk milling, promises to further enhance the sustainability and cost-effectiveness of this essential pretreatment technique. Milling is the most prevalent physical pretreatment method and dominates the scene thanks to its versatility and effectiveness [29]. However, innovative approaches like microwave irradiation and extrusion are emerging as powerful alternatives. Microwave pretreatment utilizes the unique properties of electromagnetic waves to disrupt the lignocellulosic structure, often achieving high sugar yields in a short timeframe [30]. Extrusion, combined with high temperature and shear forces, effectively breaks down the recalcitrance of lignocellulose, leading to enhanced enzymatic digestibility [31]. Both methods offer distinct advantages, with microwave pretreatment boosting energy efficiency and reduced inhibitor formation, while extrusion excels in its ability to handle diverse biomass types and produce high-quality sugars.

Ultrasonication, harnessing the power of sound waves, stands out as a promising frontier in physical pretreatment. By creating cavitation bubbles that collapse and generate reactive radicals (H and OH) in the medium to disrupt the lignocellulosic structure, it efficiently increases surface area and digestibility [32]. While factors like frequency, duration, and power need careful optimization to achieve optimal results, the potential benefits are substantial. Ultrasonication can significantly reduce hydrolysis times, leading to faster biofuel production. Its ability to pretreat various materials, including sugar beet shreds and filter paper, further underscores its versatility [33]. Although energy consumption remains a concern, ongoing research in optimizing parameters and integrating ultrasonication with other processes holds great promise for unlocking the full potential of this innovative pretreatment method.

3.2 Chemical pretreatment for enhanced material processing

3.2.1 Alkaline pretreatment

Solubilizing lignin, the recalcitrant barrier shields cellulose and hemicellulose from enzymatic digestion. The workhorses of the alkali pretreatment are sodium, potassium, calcium, and ammonium hydroxides. Sodium hydroxide (NaOH) leads the charge, cleaving the bonds between lignin and other polymers and releasing them into the solution [34]. This exposes the cellulose for enzymes to feast on and swells its structure, reducing crystallinity and increasing surface area for digestion. As if that is not enough, alkali pretreatment removes acetyl groups and uronic acids from hemicellulose, further enhancing enzymatic accessibility [35]. The evidence for the prowess of alkali pretreatment is overwhelming. Sugarcane bagasse, optimized with alkali pretreatment, yielded 82.2% more methane [36]. Rice straw shows a similar trend, treated with 1% NaOH for 3 h at room temperature, and saw a 34% boost in methane yield [37]. Ammonia-based pretreatments are also gaining traction due to their eco-friendly nature and ease of recovery. Studies on hardwood species and energy cane bagasse demonstrate their potential for sugar production [38]. Despite its effectiveness, alkali pretreatment has its Achilles' heel: alkali recovery. This adds complexity and cost to the process. Additionally, while effective for low-lignin biomass like herbaceous crops and agricultural residues, hardwoods remain a tougher nut to crack. In the end, alkali pretreatment remains a powerful tool for unlocking the potential of lignocellulosic biomass. Addressing its recovery challenges and tailoring it for diverse biomass types will further strengthen its position as a key player in the bioeconomy revolution.

3.2.2 Acid pretreatment

The acid pretreatment process utilizes the acidic properties of various chemicals to break down the complex structure of lignocellulosic biomass, primarily targeting and break the glycosidic bonds via S_N1 reaction that hold cellulose and hemicellulose together [39]. Both inorganic and organic acids are employed, each offering distinct advantages and disadvantages. Mineral acids like sulfuric, phosphoric, nitric, and hydrochloric are commonly used for their effectiveness and low cost. However, they can be highly corrosive and generate inhibitory byproducts during pretreatment [40]. Organic acids like formic, maleic, and oxalic acid are gentler on the biomass and produce fewer inhibitors, but their efficacy might be lower, and their cost can be higher [41]. Acid pretreatments can be tailored depending on the specific LB and desired outcome. Concentrated acids at range 30–70% at low temperatures lower than 100°C achieve rapid sugar conversion but are associated with high costs and environmental concerns. Dilute acids at range 0.1–10% at high temperatures from100 to 250°C are more economical and environmentally friendly but require longer processing times. One major challenge of acid pretreatment is the formation of inhibitory compounds like furfurals and phenolic acids, which hinder subsequent enzymatic hydrolysis and fermentation [42]. Chemical and biological detoxification methods are employed to mitigate this issue. Chemical approaches involve using adsorbents like activated charcoal or enzymatic treatments [43]. Biological methods utilize microbes to convert inhibitors into less harmful substances [44]. The optimal acid and pretreatment conditions depend on several factors, including the type of LB, desired sugar yield, inhibitor formation, and cost-effectiveness. Studies have shown that dilute sulfuric acid is often the most effective and economical choice for pretreating various LBs [45]. However, research on organic acids, surfactants, and targeted pretreatments of specific LB components holds promise for further improvements in efficiency and sustainability [45, 46]. Acid pretreatment plays a crucial role in unlocking the potential of LBs for biofuel production. Careful optimization of acid type, pretreatment conditions, and downstream detoxification processes is essential for maximizing sugar yield and achieving a cost-effective and environmentally friendly biofuel production system.

The quest for sustainable biofuels from lignocellulosic biomass hinges on efficient pretreatment, breaking down the complex structure for enzymatic conversion. While traditional methods often rely on harsh chemicals and energy-intensive processes, three promising alternatives emerge, including ionic liquids (ILs), organosolv, and deep eutectic solvents (DES). Under acidic conditions, benzyl cation, an active intermediate of lignin, is produced during the separation process, and its condensation will deepen the color of lignin [47]. A large number of literatures have reported that alcohols with high boiling points, such as diols or triols, can effectively protect the benzyl groups of lignin in the acid-catalyzed treatment of hardwood and herbaceous lignocellulose, thus effectively separating light-colored and low condensation lignin for further application [48–50]. So that why, recently, DES was synthesized with choline chloride, high-boiling-point alcohol (1,4-butanediol glycerin, propylene glycol, and ethylene glycol), and Lewis acid ($AlCl_3$, $FeCl_3$, $ZnCl_2$, and $CuCl_2$) for acid pretreatment of softwood to extract lignin.

3.2.3 Ionic liquids

Introduced in 2002, ILs offer fascinating possibilities. These designer solvents, with custom-tailored cations and anions, can dissolve cellulose and lignin, enhancing enzymatic digestibility [47]. Their adjustable properties, like low volatility and reusability, make them "green" champion candidate options. Imidazolium-based ILs were effective in improving sugar yield from LBs up to 97.7%. However, cost and potential enzyme toxicity remain hurdles [48, 49]. On the other hand, low-cost triethylammonium hydrogen sulfate, showcasing high lignin (up to 85%) and hemicellulose solubilization with efficient recovery, with a 75% deduction pretreatment cost, offers promising insights for future development [50]. Acid catalysts, such as H_3PO_4 and H_2SO_4, are commonly used in ionic liquids to break down the β-O-4 linkage (**Figure 4**) [51, 52].

3.2.4 Organosolv

This method utilizes organic solvents like acetic acid, acetone, dimethyl sulfoxide (DMSO), ethanol, formic acid, methanol, or tetrahydrofuran mixed with inorganic acid catalysts to break down lignocellulose biomass internal bonds (lignin and/or hemicellulose), leaving behind pure cellulose [53]. Its advantages include easy solvent recovery and high-quality lignin byproduct generation. However, the process is energy-intensive, and the high cost and volatility of organic solvents pose concerns. Research on efficient solvent recovery and exploration of less volatile alternatives is crucial for organosolv's long-term success. Ethanol organosolv mixed with 2-naphthol-7-sulfonate improved the lignin selective expelling with an 85% cellulose hydrolysis yield. High selectivity for lignin removal can be beneficial for lignin polymer utilization [54].

3.2.5 Deep eutectic solvents: a new Dawn in pretreatment

DES, a newcomer in the field, offers a compelling alternative. These mixtures of two or three components, often interlinked by hydrogen bonds, are liquid at low temperatures and possess properties similar to ILs. Chlorine chloride in the lactic acid by the ratio of 1:10, at 130°C for 4 h, successfully boosted the lignin recovery up to 72.3% with high purity at 92.5% from the lignin removal rate of 81.2% [55]. Their key advantages lie in their lower cost, biodegradability, and ease of synthesis. These advantages make DES a promising contender for sustainable LB processing [56].

3.2.6 The road ahead: a balancing act

Each method presents unique strengths and weaknesses. ILs offer tailored properties and high efficiency but face cost and toxicity challenges. Organosolv, a familiar option, requires efficient solvent recovery and cost reduction. DES, the new kid on the block, holds promise due to its sustainability, but further research is needed to optimize its performance. Ultimately, the ideal pretreatment method for LB will strike a delicate balance between efficiency, cost, and environmental impact.

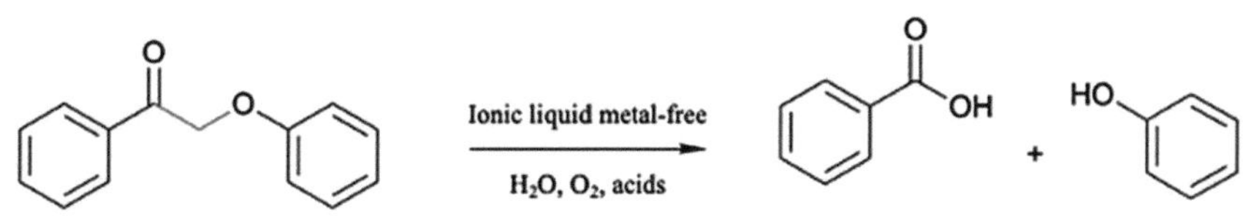

Figure 4.
Oxidation depolymerization of lignin via ionic liquids.

Continued research and development in these areas will pave the way for a future where sustainable biofuels from LBs become a reality.

4. Lignin depolymerization

4.1 Chemo-catalytic depolymerization

The burgeoning field of chemo-catalytic depolymerization offers a transformative approach to resource recovery by breaking down complex polymers into their valuable monomeric constituents. This exciting technology transcends the limitations of traditional depolymerization methods, often plagued by harsh conditions, incomplete conversion, and undesired byproducts. By harnessing the synergistic interplay of chemical reagents and tailored catalysts, chemo-catalytic depolymerization unlocks a treasure trove of possibilities [57]. Imagine lignin waste from the bioethanol or paper industries being deconstructed at the molecular level, its hydrocarbon chains cleaved with exquisite precision to yield pristine monomers. These recovered building blocks can then be seamlessly reintegrated into novel polymer production, forming a closed-loop system that dramatically reduces reliance on fossil fuels and minimizes environmental footprint.

The chemo-catalytic depolymerization lies in its exquisite selectivity. Catalysts act as molecular matchmakers, guiding the depolymerization process toward desired products while minimizing unwanted side reactions [58]. This level of control paves the way for the recovery of not just bulk monomers, but also specific high-value chemicals trapped within complex polymers. This opens doors to exciting new avenues in fields ranging from green chemistry to biorefinery. However, challenges remain. Designing efficient and cost-effective catalysts for diverse polymer feedstocks is an ongoing pursuit. Additionally, optimizing reaction conditions and overcoming potential scalability hurdles are crucial steps toward translating this promising technology into real-world applications. Despite these challenges, the potential of chemo-catalytic depolymerization to revolutionize resource utilization is undeniable. By harnessing the power of chemistry and catalysis, researchers can unlock a future where lignin waste becomes a valuable resource, paving the way for a more sustainable and circular economy. Oxidative and reduction depolymerization are well-known chemo-catalytic depolymerizations for lignin valorizations.

4.1.1 Oxidative depolymerization

The study of lignin oxidation mechanisms is crucial for optimizing its depolymerization into valuable chemicals. Oxidation depolymerization of lignin from LB is favorable for producing various valuable aromatic compounds, including aromatic derivative monomeric aldehydes to aromatic acids, and carboxylic acids. Lignin-derivative aldehydes produced by the oxidation process, namely, *p*-hydroxybenzaldehyde, vanillin, and syringaldehyde, are known to be valuable compounds that can be used as alternative bio-based chemicals (**Figure 5**) [59, 60]. The oxidation depolymerization of lignin results in the cleavage of the C−C bond, C−O−C, aromatic rings, β-O-4 bonds of lignin, and other linkages by employing the oxidants such as hydrogen peroxide (H_2O_2), oxygen (O_2), nitrobenzene, chlorine oxide, or metal oxides [61–63]. Different oxidants like H_2O_2 and O_2 initiate distinct reaction pathways via radical and electrophilic substitution in the alkaline environment. However, both pathways yield

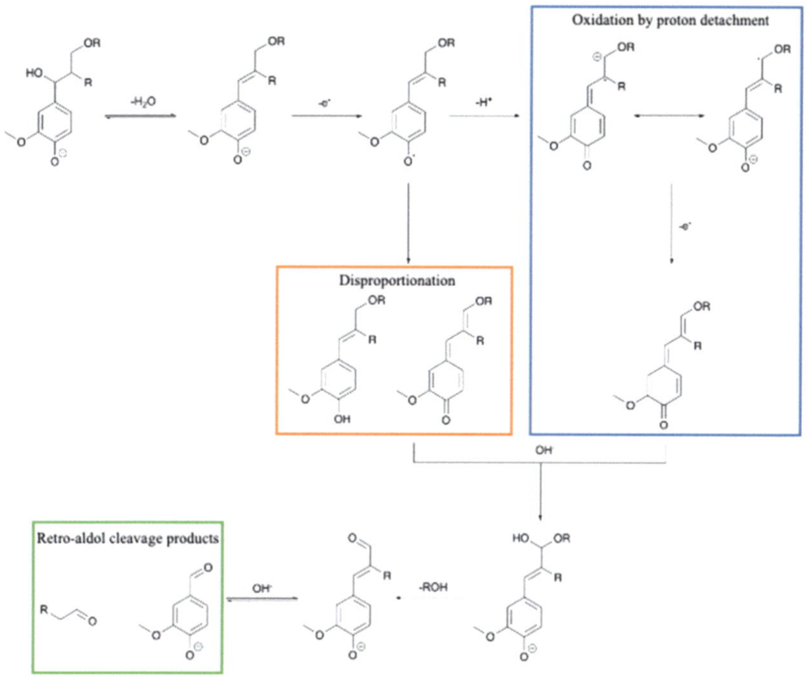

Figure 5.
Reaction mechanism for vanillin (model) via alkaline oxidative of lignin.

similar products like *p*-hydroxybenzaldehyde, syringaldehyde, and vanillin as monomeric aldehydes and its monomeric acids such as *p*-hydroxybenzoic acid, vanillic acid, and syringic acid, with alkaline conditions favoring deprotonation and phenoxy radical formation [64, 65].

The alkaline solvent is commonly used in oxidative depolymerization. In alkaline environments, phenolic compounds readily lose their protons, transforming into negatively charged phenolate anions. This deprotonation sets the stage for further oxidation, primarily through the formation of phenoxy radicals. These radicals can also arise indirectly, as peroxy radicals generated from dioxygen radicals react with phenolate anions, creating peroxide anions. These highly reactive peroxide anions then engage in an intramolecular nucleophilic attack on the carbon atom of the phenolate species, ultimately leading to the formation of dioxetane intermediates, key players in the subsequent oxidative degradation of lignin [66].

In the nitrobenzene oxidation, the reaction begins with the activation of nitrobenzene by a strong base, typically sodium hydroxide (NaOH), to form a nitrobenzene radical anion. The nitrobenzene radical anion then transfers an electron to a lignin phenolate anion, a negatively charged aromatic ring present in the lignin structure. This oxidizes the phenolate anion to a phenoxyl radical, and the nitrobenzene radical anion is converted back to neutral nitrobenzene. The phenoxyl radical undergoes a series of cleavage and rearrangement reactions, depending on the specific lignin structure. Common pathways include β-O-4 cleavage (cleavage of the Cα-Cβ bond adjacent to the phenolic oxygen, leading to fragments containing aldehyde groups), α-O-4 cleavage (cleavage of the Cβ-Cγ bond, also resulting in aldehyde-containing fragments), and side-chain cleavage, resulting in fragments containing aldehyde groups. These aldehyde fragments can further undergo oxidation and decarboxylation to form the final products, namely vanillin (from guaiacyl units in lignin) and syringaldehyde (from syringyl units in lignin). While most reaction pathways favor aldehyde formation, excessive oxidation (overoxidation) can break down the aldehyde chain further, leading to the corresponding aromatic acids [59, 67, 68]. Some intermediate hydroxy acids formed during lignin cleavage can also undergo decarboxylation to yield aromatic acids, and certain side reactions with intermediates can also form aromatic acids as minor products. Compared to oxygen, nitrobenzene showed significant results for better production of lignin monomeric aldehydes and acids. In the 10% of NaOH, nitrobenzene significantly outperformed oxygen oxidation [69]. The influence of Cu (metal oxide) in lanthanum cobaltite ($LaCoO_3$) catalysts on product yield. Increasing Cu content generates more active sites, boosting p-hydroxybenzaldehyde, vanillin, and syringaldehyde production [70]. The use of polyoxometalate catalysts. While $H_3PMo_{12}O_{40}$ in N_2 yields low products due to catalyst reduction, adding methanol as a scavenger improves conversion by preventing re-polymerization. Their study with 80% methanolic water and recycled catalysts shows up to 7 wt% combined vanillin and methyl vanillate yield [71]. In terms of specific oxidation, reactions can be employed in the production of biopolymers, such as adhesives, films, and coatings. The specific oxidation of biopolymers can be achieved using various oxidizing agents, such as periodate, hypochlorite, or nitroxyl radicals, under controlled conditions to target-specific functional groups within the polymer structure (like -OH, C−C cleavages). One prominent example is the oxidation of polysaccharides like starch to produce biopolymer derivatives with desirable properties. Soy flour adhesives were reported to be produced from insoluble soy flour via sodium periodate oxidation. Not only starch but also other sources like protein, tannin, and lignin can be used as raw materials to generate valuable products [63, 72, 73].

4.1.2 Reduction depolymerization

The reductive depolymerization of lignin is a method that harnesses hydrogen's potential by selectively snipping its C−O and C−C bonds through a medley of hydrogenation, hydrogenolysis, and hydrodeoxygenation reactions. Biofuel containing high oxygen can cause chemical instability and high viscosity, which were found to be disadvantages for industrial applications. Subsequently, the reductive depolymerization can generate a high H/C ratio or hydrogen-to-oxygen ratio of lignin monomer products by eliminating the excess oxygen into the water via hydrodeoxygenation, this process can be beneficial for biofuel or biochemical productions. Various catalysts are used as reduction depolymerization agents, including ruthenium (Ru), palladium (Pd), gold (Au), and nickel (Ni), and the combination with carbon (C) or carbon nanotube (CNT).

Employing CNT as a support for Ru catalysts and positioning these catalysts at the interface of water and oil droplets to facilitate hydrodeoxygenation of vanillin to *p*-creosol under mild reaction conditions. The Ru/CNT catalysts exhibit outstanding activity and selectivity in deoxygenating vanillin, a representative bio-oil compound. This impressive performance paves the way for streamlining bio-oil refining by merging reaction and separation steps within the stable emulsions, thereby offering cost-effective and efficient biofuel production [74]. The Pd/$CuFe_2O_4$ catalyst exhibited high activity and selectivity for the transfer hydrodeoxygenation of vanillin to 2-methoxy-4-methylphenol (MMP) with isopropanol (IPA) at 150°C. The synergistic interaction between Pd NPs and $CuFe_2O_4$ facilitated the efficient adsorption of vanillin and IPA, and the high dispersion of Pd NPs lowered the hydrogen abstraction barrier of IPA, resulting in a 99.4% vanillin conversion and 99.2% MMP selectivity. To produce 4-methyl cyclohexanol (4-MC) from vanillin, Ni was incorporated into Pd/$CuFe_2O_4$. The interfacial interaction between Pd and Ni over $CuFe_2O_4$ (1Pd7Ni/$CuFe_2O_4$) provided 95.8% 4-MC yield in H_2 (20 bar) at 180°C. The high activity of this catalyst was attributed to the acidity of the catalyst, the formation of Pd-Ni alloys, and the efficient adsorption of vanillin and MMP [75]. A combination of catalysts, including NiRu, NiRh, and NiPd, was found to outperform single-component catalysts in breaking down lignin bonds. The optimized NiRu catalyst, containing 85% nickel and 15% ruthenium, excels at breaking β-O-4 bonds at low temperatures and hydrogen pressures. It also efficiently converts lignin into valuable aromatic chemicals, significantly better than pure nickel and ruthenium catalysts. This superior performance stems from a synergistic effect by increased surface atoms, enhanced hydrogen and lignin activation, and reduced benzene ring hydrogenation. Similarly, NiRh and NiPd catalysts show improved activity and selectivity in breaking down lignin, suggesting broad potential for this family of catalysts [76].

4.2 Thermal depolymerization

Exploiting the complex chemistry of lignin, a promising avenue for sustainable resource utilization lies in thermal depolymerization. This process, often akin to pyrolysis, utilizes controlled heat in the absence of oxygen to break down the intricate lignin network. Researchers can guide depolymerization toward desired outcomes by carefully manipulating temperature, pressure, and reaction time. At lower temperatures, selective cleavage of specific bonds yields valuable phenolic monomers and oligomers, paving the way for green chemicals and bioplastics. Conversely, higher temperatures promote extensive fragmentation, generating bio-oils with potential applications in fuels and lubricants. Understanding the complex interplay between reaction conditions and product distribution remains a key challenge. By employing advanced characterization techniques and computational modeling, researchers continuously refine thermal depolymerization protocols, aiming to unlock the full potential of this transformative technology for a sustainable future.

4.2.1 Pyrolysis and microwave

Pyrolysis offers a promising approach to unlocking the hidden energy potential within LB and residual wastes like lignin. The method involves high temperatures ranging from 250 to 900°C and an inert atmosphere in the lack or limited level of oxygen as an oxidizing factor and transforms the biomass into vapor form of several organic compounds, such as alcohols, benzene, and phenols that afterward processes into valuable products like bio-oil by condensation, bio-char (bio-charcoal) as a solid residue, and leftover gases into syngas [77].

Temperature and duration reign in shaping the products of pyrolysis reactions. During slow pyrolysis, low temperatures and extended holding times favor ample syngas production that can be used as cooking gas and charcoal formation, ideal for applications like biofertilizers or adsorbents. Conversely, fast pyrolysis, with its high temperatures (800°C up to 900°C) and fleeting holding times (less than 2 s), boosts bio-oil yields, enriched with aromatic compounds from lignin breakdown [78]. However, optimizing this fast pyrolysis requires careful tuning for temperature, holding time, and flow rate, all of which influence the product distribution, highlighting the intricate interplay between these variables. Aside from temperature and time, the feedstock type also plays an important role in product distribution.

Microwave implementation has gained popularity for pyrolysis due to its rapid heating rates and uniform heating at the molecular level in comparison with conventional pyrolysis. Pyrolyzing sewage sludge at 490°C with 400 W of microwave power produced a considerably higher bio-oil content (49.7 wt%) compared to conventional heating at 500°C (36.7 wt%) [79, 80]. The fast-holding time of oil in the microwave avoids secondary cracking to gaseous products, resulting in a higher bio-oil yield [81]. Despite the advantages of microwave-assisted pyrolysis, challenges remain in its practical application, particularly for lignin depolymerization. The product selectivity in pyrolysis tends to be low, and separating the various products poses a significant challenge. This limitation in product separation could hinder the widespread adoption of microwave-assisted pyrolysis for lignin depolymerization. Researchers are addressing these challenges to improve the efficiency and feasibility of this innovative pyrolytic approach.

4.2.2 Gasification

Gasification is an alternative thermal depolymerization of LB into synthetic fuel, such as methanol, dimethyl ether, and paraffin, through the Fischer-Tropsch mechanism [82]. The process is conducted at extremely high temperatures above 800°C while maintaining the presence of air or oxygen as an oxidizing agent [83]. The sequence of the gasification process includes oxidation by combustion that forms CO and CO_2 (from char) and water (from H_2 of syngas), drying steps to eliminate the moisture in LB, then pyrolysis decomposition, and lastly, reduction steps that produce CO and H_2 (boudouard reaction) also CH_4 and H_2 (methanation reaction). The production of CO and H_2 leads to the Fischer-Tropsch process [84]. Decrement of the carbon-to-hydrogen ratio through gasification enhances the product's calorific value and simultaneously generates hydrogen, which is important for the reductive depolymerization of lignin into valuable aromatics [85].

4.3 Photocatalytic depolymerization

Photocatalytic depolymerization uses light and a photocatalyst to break down lignin into smaller molecules. The photocatalyst absorbs light, creating excited electrons and holes that initiate chemical reactions. These reactions can cleave lignin bonds, leading to the formation of various products like monomers, oligomers, and aromatic chemicals (**Figure 6**) [86].

Compared to traditional methods like thermochemical or biological depolymerization, photocatalysis offers several advantages, such as mild reaction conditions, lignin can be depolymerized at lower temperatures and pressures, minimizing energy consumption and potential degradation of valuable products. Second, the choice of photocatalyst and reaction conditions can potentially target-specific lignin bonds, leading to the desired products. Lastly, the sustainability aspect. Photocatalysis uses light that mostly comes from unlimited resources like sunlight and readily available resources, making it a potentially environmentally friendly approach [87]. Despite its environmental merits, some challenges arise, including low product yields, due to current methods still struggling with efficiently converting lignin into valuable products. The second is the catalyst stability, and photocatalysts can deactivate or degrade over time, requiring design improvements. The last one is reactor design for

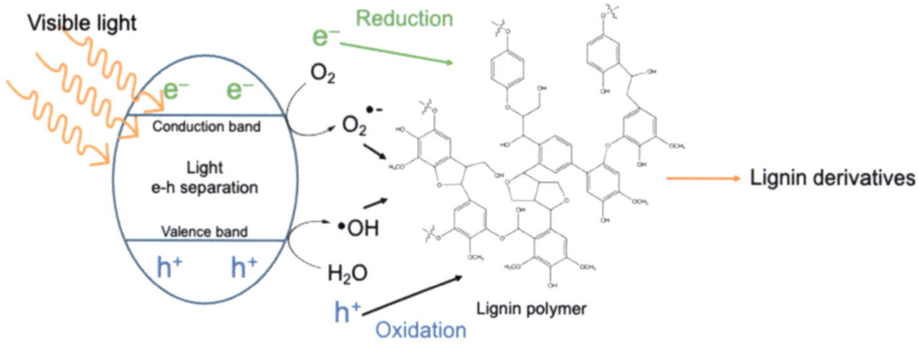

Figure 6.
In the heterogeneous photocatalytic conversion of lignin, the black arrow is an oxidation-related process, while the green arrows represent reduction-related steps within the mechanism.

optimizing light utilization and mass transfer within the reactor, which is crucial for efficient depolymerization and scaling up for industrial implementations that require cost-effective materials and processes [88].

4.4 Electrochemical depolymerization

Electrochemical depolymerization has gained significant interest as a method to enhance the conversion of lignin from LB. This process is advantageous because it can be conducted at room temperature and in a normal atmosphere [89, 90]. The electrochemical depolymerization of lignin offers many notable advantages compared to thermochemical and chemocatalytic methods of lignin degradation. Firstly, the process does not require specific amounts of oxidants, increasing efficiency and reducing environmental impact. Secondly, environmentally friendly byproducts are produced, with water being a major part of the process [91]. Thirdly, gentle reaction conditions without requiring delicate chemicals are used [92]. Fourthly, the lignin degradation process can be precisely controlled by adjusting the electrode potential or current intensity, resulting in high selectivity for desired outcomes [93]. Additionally, combining electrochemical lignin upgrade with renewable energy harvesting is a promising approach to achieving carbon neutrality [94–96]. Electrochemical lignin depolymerization aims to selectively disassemble lignin into specific chemical intermediates or monomers. By employing particular conditions and catalysts, it is feasible to produce useful molecules such as phenols, vanillin, syringaldehyde, and other aromatic substances selectively.

The primary strategies used for lignin depolymerization based on electrocatalytic oxidation consist of four modes, i.e., mediated electrooxidation, electroenzymatic oxidation, photoelectrochemical oxidation, and direct electrooxidation (**Figure 4**).

In mediated electrooxidation, a redox-active mediator facilitates electron transfer between the anode and lignin polymers (as depicted by type I in **Figure** 7). The redox-active mediator undergoes electrochemical transformation at the anode, which is oxidized. Subsequently, the oxidized mediator reacts chemically with the lignin polymer present in the electrolyte, leading to the breakdown of the lignin polymer and the formation of its oxidized products. Following this reaction, the mediator returns to its reduced state, completing the cycle. One of the advantages of using soluble catalysts in this mediated electrooxidation process is that the lignin substrate can be present in the solution as a slurry [97]. This allows for efficient lignin processing under suitable conditions, enhancing the overall efficiency of the electrochemical oxidation process.

Under electroenzymatic oxidation conditions, a lignin depolymerization enzyme such as LiP and MnP in the solution is activated by an electrochemically generated oxidant, often H_2O_2. Subsequently, the activated enzyme interacts with lignin to produce oxidized lignin products (as depicted by type II in **Figure 7**) [98]. In this mode, lignin oxidation typically undergoes a cathodic process. Researchers have developed LiP and MnP-inspired Mn-Schiff complexes as mediators for the electroenzymatic oxidation of lignin [90]. These complexes play a role similar to redox-active mediators in mediated electrooxidation, facilitating electron transfer between the enzyme and lignin substrates, thus promoting the depolymerization of lignin.

In photoelectrochemical systems for lignin oxidation, the photogenerated hole within a photoanode is utilized to oxidize the redox mediator, thereby initiating the lignin oxidation reaction (as illustrated by type III in **Figure 7**) [99]. A notable example of such a system is reported by Li et al. [100], where they introduced a photoelectrochemical PINO-catalyzed lignin oxidation method. In this method, a ruthenium polypyridyl photocatalyst (RuC) is immobilized on the surface of a TiO_2 photoanode. Under visible light irradiation, the excited state photocatalyst (RuC*) donates one electron to the TiO_2 semiconductor, generating an oxidized state photocatalyst (RuC+). The NHPI/2,6-lutidine co-catalyst is then sufficiently oxidized to PINO radical by RuC+ in the electrolyte, initiating the lignin oxidation reaction.

In heterogeneous direct electrocatalytic oxidation mode, a solid-state catalyst serves as an electrode or is immobilized on the electrode surface, enabling direct electrochemical oxidation of the lignin polymer on the electrode surface (as depicted by type IV in **Figure 7**). Metal oxides have shown significant electrocatalytic activity for the oxidation of organic compounds [101–107]. Numerous studies have achieved successful heterogeneous electrocatalytic oxidation of lignin using transition metals, metal oxides, or alloys as the anode material. While theoretically highly efficient, several issues limit the effectiveness of lignin depolymerization via this method. One major issue is anode deactivation, primarily due to the formation of a passivation layer on the anode surface during lignin electrolysis, hindering active site accessibility and increasing internal resistance [102, 108, 109].

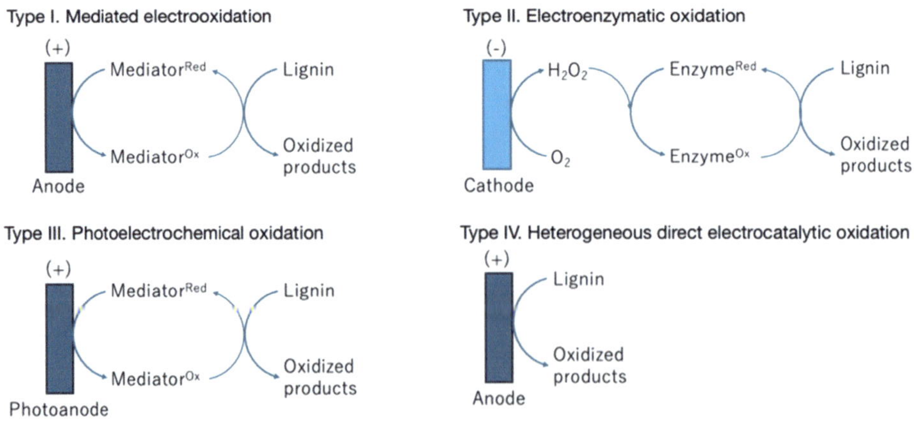

Figure 7.
Classification of electrocatalytic lignin depolymerization based on oxidation.

Electrochemical lignin depolymerization has obstacles concerning reaction selectivity, energy efficiency, and cost-effectiveness, notwithstanding its potential. Scientists are continuously improving the method and creating effective catalysts to enhance its feasibility for large-scale utilization of lignin.

5. Valorization of lignin by biological platforms

The past decade has witnessed a surge in microbial production of biofuels and valuable products like lipids, vanillin, biopolymers, and other valuable lignin monomeric aromatic compounds (aldehydes, acids, and alcohols) directly from lignin and its derivatives. This exciting development hinges on the efficient breakdown of lignin, the initial step in biofuel and biomaterial synthesis. Enzymatic degradation by microbes, particularly fungi, bacteria, and yeasts, offers a promising eco-friendly approach to unlocking lignin's potential. Remarkably, these microbes produce a diverse array of lignin-degrading enzymes, including laccases, peroxidases, and dye-decolorizing peroxidases, paving the way for further advancements in lignin utilization [110]. Recent advancements in screening, strain development, and enzyme utilization strategies have further amplified their effectiveness, paving the way for a new era of efficient lignin valorization. As research delves deeper into the lignin valorization by microbial platforms, the possibilities for sustainable biofuel and biomaterial production from this once recalcitrant resource become increasingly limitless.

5.1 Lignin valorization by fungi

As nature decomposers, white-rot fungi (mainly from Basidiomycota) hold immense potential beyond their well-known ability to break down lignin [111]. White-rot fungi outperformed brown-rot fungi and soft-rot fungi in terms of capability to degrade lignin. Their versatility and efficiency make them valuable tools for a range of applications, from environmental cleanup to biofuel production, which lies in their ability to produce lignin-degrading enzymes, including laccases, lignin peroxidases (LiP), manganese peroxidases (MnP), and versatile peroxidases (VP) (**Figure 8**) [112]. Lignin degradation through laccases occurs through either oxidation or the acceptance of an oxygen molecule as an electron donor. LiP exhibits similarities with other peroxidases regarding their catalytic and oxidative [5]. The peroxidase enzymes can catalyze lignin encompassing phenolic and non-phenolic varieties, using H_2O_2. This catalytic activity is attributed to the presence of an iron protoporphyrin IX for each protein molecule in these enzymes. Specifically, LiP typically employs long-range electron transfer mechanisms to oxidize lignin compounds. In contrast, MnP utilizes Mn (III) oxidation as a diffusible intermediate for its catalytic process. Versatile peroxidases (VP), on the other hand, operate through a combination of both electron transfer and diffusible intermediate in their mode of action [113].

In vitro applications leverage these enzymes to depolymerize lignin and its derivatives, presenting a promising alternative to directed-cell lignin conversion. The adoption of in vitro enzymatic conversion offers several advantages, such as enhancing substrate and enzyme interaction. This approach also holds the potential to reduce cultivation time and address ATP/NAD(P)H imbalance, showcasing its versatility and efficiency in lignin transformation processes [114].

Among other white-rot fungi, *Phanerochaete chrysosporium* is known as the best lignin-degrading fungi for LB by using MnP and LiP through a non-specific oxidative

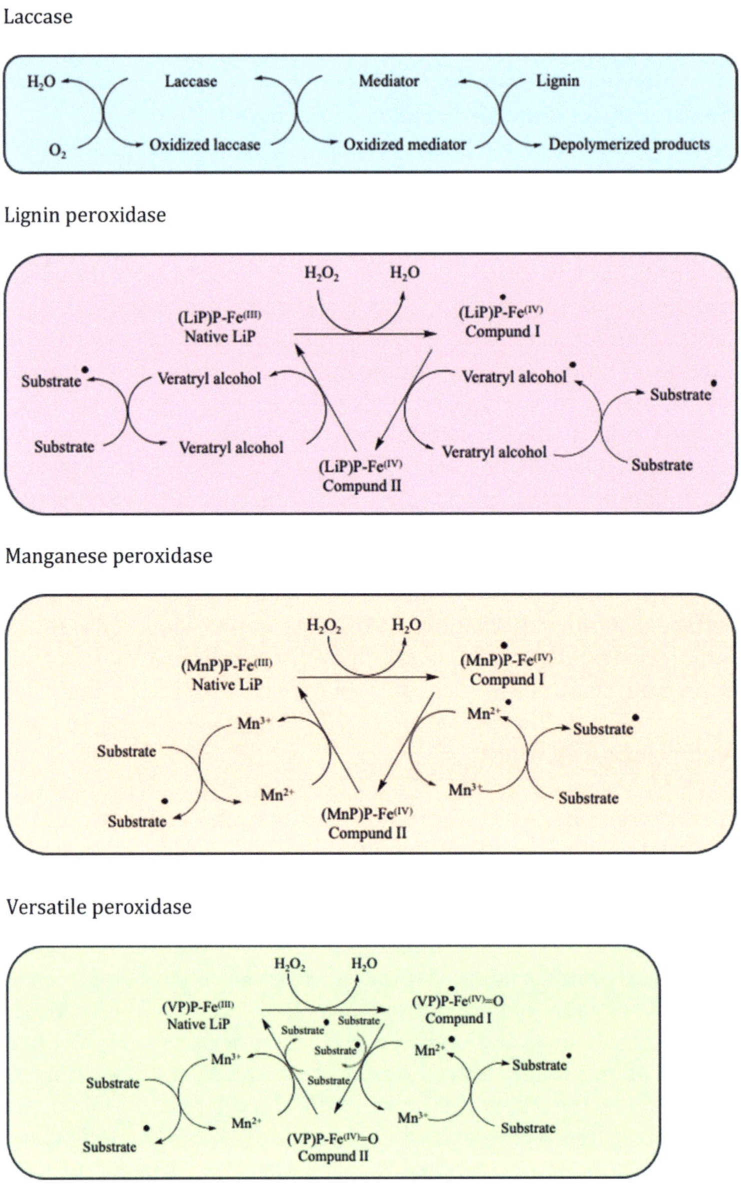

Figure 8.
Catalytic mechanism of enzymatic depolymerization.

mechanism [115, 116]. Other white-rot fungi that are able to utilize lignin, hemicellulose, and cellulose are *Irpex lacteusare*, *Heterobasidion annosum*, *Trametes versicolor*, and including *P. chrysosporium* [111]. Due to their impressiveness to lignin breakdown, the low-molecular-weight products generated by this process hold untapped potential. These lignin molecules can be transformed into valuable renewable chemicals, like vanillic acid and ferulic acid, used in everything from pharmaceuticals to cosmetics [117]. Some white-rot fungi can do more than just break down lignin, and they can actually use it as a source of energy. By converting lignin into fatty acids, like acetic acid, they unlock a new source of biofuel production.

Even though some species of white-rot fungi do not have the ability to degrade cellulose, only lignin and hemicellulose, such as *Ceriporipsis subvermispora*, *Ganoderma australe*, *Phellinus pini*, and *Phlebia tremellosa* [118]. These types of white-rot fungi can enhance the papermaking process by selectively removing lignin, leaving behind strong, pure cellulose fibers. This bio-pulping technique leads to stronger, brighter paper with reduced environmental impact. It is beneficial to minimize the use of deforestation and harsh chemicals for sustainable paper production. Another exciting avenue for lignin valorization lies in demethylation. By removing methyl groups from lignin, fungi increase its reactivity, making it more suitable for creating lignin-based polymers [119]. This opens doors for developing new bioplastics and other sustainable materials, reducing our reliance on fossil fuels.

In conclusion, white-rot fungi are not just lignin-munching machines; they are versatile biocatalysts with the potential to revolutionize various industries. From environmental cleanup to biofuel production, these remarkable organisms offer a sustainable and efficient solution for a range of challenges. As we continue to unlock their potential, the future looks promising for white-rot fungi to become the impactful agents of a greener and cleaner world.

5.2 Lignin valorization by bacteria

While bacterial lignin degradation efficiency lags behind fungi, their versatility for adaptation in diverse environments is beneficial. *Actinobacteria*, *Firmicutes*, and *Proteobacteria* dominate this field, thriving in soil, gut, wastewater, and wood, even thriving in the absence of oxygen [120]. This broad ecological range makes them attractive candidates for lignin bioremediation in various settings. Actinobacteria like Streptomyces and Rhodococcus, as aerobic bacteria, stand out in lignin utilization. *Streptomyces* sp., such as S. *viridosporus* and *S. setonii,* efficiently depolymerizes lignin units, removing up to 45% from wood in 90 days [121]. *Rhodococcus* like *R. jostii* RHA1, known for its exceptional tolerance to toxic intermediates, degrades wheat straw and kraft lignin into valuable products like vanillin and carboxylic acids [122, 123]. Proteobacteria like *Pandoraea* sp. and *Pseudomonas putida* also join the fray, catabolizing lignin and producing various aromatic compounds. Bacteria thrive using oxygen to break down lignin.

While aerobic bacteria have some merits for lignin utilization, anaerobic bacteria have the ability to convert lignin into carbon dioxide and methane [124, 125]. Bacteria prefer lignin with more methoxyl groups, using them as a docking point for their enzymatic attack. The process unfolds in three acts: methoxyl removal, aromatic ring cleavage, and methane formation. Methanogenic microbes turn the lignin-derived aromatics into the final act of methane production. The paper and pulp industry even hosts its anaerobic decomposers like *Acetoanaerobium* sp., which oxidizes kraft lignin and produces valuable aromatic compounds like ferulic and syringic acid [126].

From the sun-drenched surface to the oxygen-deprived depths, these remarkable bacteria reveal a hidden world of lignin degradation, with each act in their play offering valuable insights for biofuel production, environmental remediation, and sustainable resource utilization.

5.3 Lignin valorization by yeast

As a part of microorganisms like fungi and bacteria, the potential of yeasts for biofuel production from lignin monomeric compounds is promising. Olegianous yeast like *Trichosporon cutaneum* and *T. oleaginosus* demonstrate remarkable tolerance and biodegradation of phenolic aldehydes, metabolizing them into lipids via diverse pathways. *T. cutaneum* utilizes 4-hydroxybenzaldehyde for lipid production (16.6%) [127], while *T. oleaginosus* thrives on 4-hydroxybenzoic acid, achieving a 69.5% lipid yield [128]. Further research on the correlation between specific yeast–lignin interactions and lipid production will be crucial for optimizing industrial-scale biofuel production.

While lipid production from lignin holds significant promise, exploring other valuable products derived from yeast metabolism unlocks the full potential of lignin valorization. For instance, *T. cutaneum* and *T. guehoae* reported converting lignin aldehydes like syringaldehyde and vanillin into their corresponding acids and alcohols. On the other hand, *Lipomyces starkeyi* plays its role in the production of valuable lignin monomeric alcohols such as 4-hydroxybenzyl alcohol, vanillyl alcohol, and syringyl alcohol from its aldehyde derivative via aldehyde-ketone reductase (AKR) conversion [129]. This opens avenues for pharmaceutical and chemical applications beyond biofuels, further enhancing the economic viability of lignin utilization. Further research in this area can lead to the development of engineered yeasts with enhanced bioconversion efficiency and tailored product output, unlocking a sustainable future for lignin valorization.

6. Conclusion

Lignin valorization presents a critical challenge and exciting opportunity for the biorefinery industry. The future of the biorefinery landscape hinges on effectively valorizing lignin. This intricate biopolymer offers a treasure trove of possibilities, from replacing petroleum-derived chemicals to creating novel biomaterials. Unlocking its full potential requires addressing challenges like optimizing extraction and conversion processes while ensuring economic viability. We can transform lignin from a mere byproduct into a driving force for a greener and more sustainable future through a concerted effort involving researchers, industry leaders, and policymakers.

Acknowledgements

This work was partially supported by the New Energy and Industrial Technology Development Organization (NEDO) Feasibility Study Program, Japan, the International Joint Program, Science and Technology Research Partnership for Sustainable Development (SATREPS) from the Japan Science and Technology Agency, and the Japan International Cooperation Agency (JST and JICA).

Authorship contribution statement

Conceptualization, F.J.N.P., and P.K.; data curation, F.J.N.P.; visualization, F.J.N.P. and P.K.; writing—original draft, F.J.N.P. and P.K.; supervision, P.K., C.O., and A.K. All authors have read and agreed to the published version of the manuscript.

Conflict of interest

The authors declare no conflict of interest.

DOI: http://dx.doi.org/10.5772/intechopen.1005166

Author details

Filemon Jalu Nusantara Putra, Prihardi Kahar*, Chiaki Ogino and Akihiko Kondo
Graduate School of Engineering, Kobe University, Kobe, Japan

*Address all correspondence to: pri@port.kobe-u.ac.jp

References

[1] Holechek JL, Geli HME, Sawalhah MN, Valdez R. A global assessment: Can renewable energy replace fossil fuels by 2050? Sustainability. 2022;**14**:4792. DOI: 10.3390/su14084792

[2] York R. Do alternative energy sources displace fossil fuels? Nature Climate Change. 2012;**2**:441-443. DOI: 10.1038/nclimate1451

[3] Singh N, Singhania RR, Nigam PS, Dong C-D, Patel AK, Puri M. Global status of lignocellulosic biorefinery: Challenges and perspectives. Bioresource Technology. 2022;**344**:126415. DOI: 10.1016/j.biortech.2021.126415

[4] Matsushita Y. Conversion of technical lignins to functional materials with retained polymeric properties. Journal of Wood Science. 2015;**61**:230-250. DOI: 10.1007/s10086-015-1470-2

[5] Putra FJN, Kahar P, Kondo A, Ogino C. Valorization of lignin and its derivatives using yeast. PRO. 2022;**10**:2004. DOI: 10.3390/pr10102004

[6] Wang H, Pu Y, Ragauskas A, Yang B. From lignin to valuable products–strategies, challenges, and prospects. Bioresource Technology. 2019;**271**:449-461. DOI: 10.1016/j.biortech.2018.09.072

[7] Zhang C. Lignocellulosic Ethanol: Technology and Economics. Alcohol Fuels - Current Technologies and Future Prospect. London, UK: IntechOpen; 2020. DOI: 10.5772/intechopen.86701

[8] Mankar AR, Modak A, Pant KK. Recent advances in the valorization of lignin: A key focus on pretreatment, characterization, and catalytic depolymerization strategies for future biorefineries. Advanced Sustainable Systems. 2022;**6**:2100299. DOI: 10.1002/adsu.202100299

[9] Kenny JK, Medlin JW, Beckham GT. Quantification of phenolic hydroxyl groups in lignin via ^{19}F NMR spectroscopy. ACS Sustainable Chemistry & Engineering. 2023;**11**:5644-5655. DOI: 10.1021/acssuschemeng.3c00115

[10] Shakeel U, Li X, Wang B, Geng F, Rehman MSU, Zhang K, et al. Structural characterizations of lignins extracted under same severity using different acids. International Journal of Biological Macromolecules. 2022;**194**:204-212. DOI: 10.1016/j.ijbiomac.2021.11.171

[11] Ponomarenko J, Lauberts M, Dizhbite T, Lauberte L, Jurkjane V, Telysheva G. Antioxidant activity of various lignins and lignin-related phenylpropanoid units with high and low molecular weight. Holzforschung. 2015;**69**:795-805. DOI: 10.1515/hf-2014-0280

[12] Sun Z, Fridrich B, de Santi A, Elangovan S, Barta K. Bright side of lignin depolymerization: Toward new platform chemicals. Chemical Reviews. 2018;**118**:614-678. DOI: 10.1021/acs.chemrev.7b00588

[13] Upton BM, Kasko AM. Strategies for the conversion of lignin to high-value polymeric materials: Review and perspective. Chemical Reviews. 2016;**116**:2275-2306. DOI: 10.1021/acs.chemrev.5b00345

[14] Li C, Zhao X, Wang A, Huber GW, Zhang T. Catalytic transformation of lignin for the production of chemicals and fuels. Chemical Reviews.

2015;**115**:11559-11624. DOI: 10.1021/acs.chemrev.5b00155

[15] Pu Y, Cao S, Ragauskas AJ. Application of quantitative 31P NMR in biomass lignin and biofuel precursors characterization. Energy & Environmental Science. 2011;**4**:3154. DOI: 10.1039/c1ee01201k

[16] Houfani AA, Anders N, Spiess AC, Baldrian P, Benallaoua S. Insights from enzymatic degradation of cellulose and hemicellulose to fermentable sugars– A review. Biomass and Bioenergy. 2020;**134**:105481. DOI: 10.1016/j.biombioe.2020.105481

[17] Yuan Y, Jiang B, Chen H, Wu W, Wu S, Jin Y, et al. Recent advances in understanding the effects of lignin structural characteristics on enzymatic hydrolysis. Biotechnology for Biofuels. 2021;**14**:205. DOI: 10.1186/s13068-021-02054-1

[18] Lourenço A, Pereira H. Compositional Variability of Lignin in Biomass. Lignin - Trends and Applications. London, UK: InTech; 2018. DOI: 10.5772/intechopen.71208

[19] Anderson EM, Stone ML, Katahira R, Reed M, Muchero W, Ramirez KJ, et al. Differences in S/G ratio in natural poplar variants do not predict catalytic depolymerization monomer yields. Nature Communications. 2019;**10**:2033. DOI: 10.1038/s41467-019-09986-1

[20] Yoo CG, Dumitrache A, Muchero W, Natzke J, Akinosho H, Li M, et al. Significance of lignin S/G ratio in biomass recalcitrance of *Populus trichocarpa* variants for bioethanol production. ACS Sustainable Chemistry & Engineering. 2018;**6**:2162-2168. DOI: 10.1021/acssuschemeng.7b03586

[21] Meng X, Parikh A, Seemala B, Kumar R, Pu Y, Wyman CE, et al. Characterization of fractional cuts of co-solvent enhanced lignocellulosic fractionation lignin isolated by sequential precipitation. Bioresource Technology. 2019;**272**:202-208. DOI: 10.1016/j.biortech.2018.09.130

[22] Shi Z, Xu G, Deng J, Dong M, Murugadoss V, Liu C, et al. Structural characterization of lignin from *D. Sinicus* by FTIR and NMR techniques. Green Chemistry Letters and Reviews. 2019;**12**:235-243. DOI: 10.1080/17518253.2019.1627428

[23] Alves CT, Peters MA, Onwudili JA. Application of thermogravimetric analysis method for the characterisation of products from triglycerides during biodiesel production. Journal of Analytical and Applied Pyrolysis. 2022;**168**:105766. DOI: 10.1016/j.jaap.2022.105766

[24] Eugenio ME, Martín-Sampedro R, Santos JI, Wicklein B, Ibarra D. Chemical, thermal and antioxidant properties of Lignins solubilized during soda/AQ pulping of Orange and olive tree pruning residues. Molecules. 2021;**26**:3819. DOI: 10.3390/molecules26133819

[25] Meng X, Crestini C, Ben H, Hao N, Pu Y, Ragauskas AJ, et al. Determination of hydroxyl groups in biorefinery resources via quantitative 31P NMR spectroscopy. Nature Protocols. 2019;**14**:2627-2647. DOI: 10.1038/s41596-019-0191-1

[26] Gioia C, Lo Re G, Lawoko M, Berglund L. Tunable thermosetting epoxies based on fractionated and well-characterized Lignins. Journal of the American Chemical Society. 2018;**140**:4054-4061. DOI: 10.1021/jacs.7b13620

[27] Gallego-García M, Moreno AD, Manzanares P, Negro MJ, Duque A.

Recent advances on physical technologies for the pretreatment of food waste and lignocellulosic residues. Bioresource Technology. 2023;**369**:128397. DOI: 10.1016/j.biortech.2022.128397

[28] Shah AA, Seehar TH, Sharma K, Toor SS. Biomass pretreatment technologies. In: Hydrocarbon Biorefinery. Amsterdam, Netherlands: Elsevier; 2022. pp. 203-228. DOI: 10.1016/B978-0-12-823306-1.00014-5

[29] Arce C, Kratky L. Mechanical pretreatment of lignocellulosic biomass toward enzymatic/fermentative valorization. IScience. 2022;**25**:104610. DOI: 10.1016/j.isci.2022.104610

[30] Jablonowski ND, Pauly M, Dama M. Microwave assisted pretreatment of Szarvasi (Agropyron elongatum) biomass to enhance enzymatic saccharification and direct glucose production. Frontiers in Plant Science. 2022;**12**:767254. DOI: 10.3389/fpls.2021.767254

[31] Gatt E, Rigal L, Vandenbossche V. Biomass pretreatment with reactive extrusion using enzymes: A review. Industrial Crops and Products. 2018;**122**:329-339. DOI: 10.1016/j.indcrop.2018.05.069

[32] Ashokkumar M. The characterization of acoustic cavitation bubbles – An overview. Ultrasonics Sonochemistry. 2011;**18**:864-872. DOI: 10.1016/j.ultsonch.2010.11.016

[33] Flores EMM, Cravotto G, Bizzi CA, Santos D, Iop GD. Ultrasound-assisted biomass valorization to industrial interesting products: State-of-the-art, perspectives and challenges. Ultrasonics Sonochemistry. 2021;**72**:105455. DOI: 10.1016/j.ultsonch.2020.105455

[34] Wang W, Wang X, Zhang Y, Yu Q, Tan X, Zhuang X, et al. Effect of sodium hydroxide pretreatment on physicochemical changes and enzymatic hydrolysis of herbaceous and woody lignocelluloses. Industrial Crops and Products. 2020;**145**:112145. DOI: 10.1016/j.indcrop.2020.112145

[35] Singh A, Kushwaha A, Sen S, Goswami S, Katiyar S, Kumar A, et al. Recent advancement in microwave-assisted pyrolysis for biooil production. In: Waste-to-Energy Approaches Towards Zero Waste. Amsterdam, Netherlands: Elsevier; 2022. pp. 197-219. DOI: 10.1016/B978-0-323-85387-3.00014-8

[36] Shen J, Zhao C, Liu G, Chen C. Enhancing the performance on anaerobic digestion of vinegar residue by sodium hydroxide pretreatment. Waste and Biomass Valorization. 2017;**8**:1119-1126. DOI: 10.1007/s12649-016-9666-2

[37] Shetty DJ, Kshirsagar P, Tapadia-Maheshwari S, Lanjekar V, Singh SK, Dhakephalkar PK. Alkali pretreatment at ambient temperature: A promising method to enhance biomethanation of rice straw. Bioresource Technology. 2017;**226**:80-88. DOI: 10.1016/j.biortech.2016.12.003

[38] Sakuragi K, Igarashi K, Samejima M. Application of ammonia pretreatment to enable enzymatic hydrolysis of hardwood biomass. Polymer Degradation and Stability. 2018;**148**:19-25. DOI: 10.1016/j.polymdegradstab.2017.12.008

[39] Shrotri A, Kobayashi H, Fukuoka A. Catalytic Conversion of Structural Carbohydrates and Lignin to Chemicals. Advances in Catalysis. Amsterdam, Netherlands: Elsevier; 2017. pp. 59-123. DOI: 10.1016/bs.acat.2017.09.002

[40] Jönsson LJ, Martín C. Pretreatment of lignocellulose: Formation of inhibitory by-products and strategies for

minimizing their effects. Bioresource Technology. 2016;**199**:103-112. DOI: 10.1016/j.biortech.2015.10.009

[41] Qing Q, Huang M, He Y, Wang L, Zhang Y. Dilute oxalic acid pretreatment for high total sugar recovery in pretreatment and subsequent enzymatic hydrolysis. Applied Biochemistry and Biotechnology. 2015;**177**:1493-1507. DOI: 10.1007/s12010-015-1829-2

[42] Edwiges T, Roda-Serrat MC, Segovia-HernándezJG,Sánchez-RamírezE, Tronci S, Errico M. Sustainable bioalcohol production: Pretreatment, separation, and control strategies leading to sustainable processes. In: Biofuels and Biorefining. Amsterdam, Netherlands: Elsevier; 2022. pp. 41-85. DOI: 10.1016/B978-0-12-824117-2.00004-1

[43] Li J, Shi S, Tu M, Via B, Sun FF, Adhikari S. Detoxification of Organosolv-pretreated pine prehydrolysates with anion resin and cysteine for butanol fermentation. Applied Biochemistry and Biotechnology. 2018;**186**:662-680. DOI: 10.1007/s12010-018-2769-4

[44] Bhatia SK, Kim J, Song H-S, Kim HJ, Jeon J-M, Sathiyanarayanan G, et al. Microbial biodiesel production from oil palm biomass hydrolysate using marine Rhodococcus sp. YHY01. Bioresource Technology. 2017;**233**:99-109. DOI: 10.1016/j.biortech.2017.02.061

[45] Niju S, Swathika M, Balajii M. Pretreatment of lignocellulosic sugarcane leaves and tops for bioethanol production. In: Lignocellulosic Biomass to Liquid Biofuels. Amsterdam, Netherlands: Elsevier; 2020. pp. 301-324. DOI: 10.1016/B978-0-12-815936-1.00010-1

[46] Liu C-G, Li K, Wen Y, Geng B-Y, Liu Q, Lin Y-H. Bioethanol: New Opportunities for an Ancient Product. Advances in Bioenergy. Amsterdam, Netherlands: Elsevier; 2019. pp. 1-34. DOI: 10.1016/bs.aibe.2018.12.002

[47] Yoo CG, Pu Y, Ragauskas AJ. Ionic liquids: Promising green solvents for lignocellulosic biomass utilization. Current Opinion in Green and Sustainable Chemistry. 2017;**5**:5-11. DOI: 10.1016/j.cogsc.2017.03.003

[48] Xu J, Liu B, Hou H, Hu J. Pretreatment of eucalyptus with recycled ionic liquids for low-cost biorefinery. Bioresource Technology. 2017;**234**:406-414. DOI: 10.1016/j.biortech.2017.03.081

[49] Raj JJ, Magaret S, Pranesh M, Lethesh KC, Devi WC, Mutalib MIA. Dual functionalized imidazolium ionic liquids as a green solvent for extractive desulfurization of fuel oil: Toxicology and mechanistic studies. Journal of Cleaner Production. 2019;**213**:989-998. DOI: 10.1016/j.jclepro.2018.12.207

[50] Usmani Z, Sharma M, Gupta P, Karpichev Y, Gathergood N, Bhat R, et al. Ionic liquid based pretreatment of lignocellulosic biomass for enhanced bioconversion. Bioresource Technology. 2020;**304**:123003. DOI: 10.1016/j.biortech.2020.123003

[51] Binder JB, Gray MJ, White JF, Zhang ZC, Holladay JE. Reactions of lignin model compounds in ionic liquids. Biomass and Bioenergy. 2009;**33**:1122-1130. DOI: 10.1016/j.biombioe.2009.03.006

[52] Yang Y, Fan H, Song J, Meng Q, Zhou H, Wu L, et al. Free radical reaction promoted by ionic liquid: A route for metal-free oxidation depolymerization of lignin model compound and lignin. Chemical Communications. 2015;**51**:4028-4031. DOI: 10.1039/C4CC10394G

[53] Arif, Hakimi Saadon SZ, Osman NB, Yusup S. Pretreatment

of fiber-based biomass material for lignin extraction. In: Value-Chain of Biofuels. Amsterdam, Netherlands: Elsevier; 2022. pp. 105-135. DOI: 10.1016/B978-0-12-824388-6.00024-5

[54] Chu Q, Tong W, Chen J, Wu S, Jin Y, Hu J, et al. Organosolv pretreatment assisted by carbocation scavenger to mitigate surface barrier effect of lignin for improving biomass saccharification and utilization. Biotechnology for Biofuels. 2021;**14**:136. DOI: 10.1186/s13068-021-01988-w

[55] Yang L, Zheng T, Huang C, Yao J. Using deep eutectic solvent pretreatment for enhanced enzymatic saccharification and lignin utilization of masson pine. Renewable Energy. 2022;**195**:681-687. DOI: 10.1016/j.renene.2022.06.066

[56] Zdanowicz M, Wilpiszewska K, Spychaj T. Deep eutectic solvents for polysaccharides processing. A review. Carbohydrate Polymers. 2018;**200**:361-380. DOI: 10.1016/j.carbpol.2018.07.078

[57] Modak A, Mankar AR, Pant KK, Bhaumik A. Mesoporous porphyrin-silica nanocomposite as solid acid catalyst for high yield synthesis of HMF in water. Molecules. 2021;**26**:2519. DOI: 10.3390/molecules26092519

[58] Walsh DJ, Hyatt MG, Miller SA, Guironnet D. Recent trends in catalytic polymerizations. ACS Catalysis. 2019;**9**:11153-11188. DOI: 10.1021/acscatal.9b03226

[59] Tarabanko V, Tarabanko N. Catalytic oxidation of Lignins into the aromatic aldehydes: General process trends and development prospects. International Journal of Molecular Sciences. 2017;**18**:2421. DOI: 10.3390/ijms18112421

[60] Liu C, Wu S, Zhang H, Xiao R. Catalytic oxidation of lignin to valuable biomass-based platform chemicals: A review. Fuel Processing Technology. 2019;**191**:181-201. DOI: 10.1016/j.fuproc.2019.04.007

[61] Dai J, Patti AF, Saito K. Recent developments in chemical degradation of lignin: Catalytic oxidation and ionic liquids. Tetrahedron Letters. 2016;**57**:4945-4951. DOI: 10.1016/j.tetlet.2016.09.084

[62] Ouyang X, Tan Y, Qiu X. Oxidative degradation of lignin for producing monophenolic compounds. Journal of Fuel Chemistry and Technology. 2014;**42**:677-682. DOI: 10.1016/S1872-5813(14)60030-X

[63] Chen X, Xi X, Pizzi A, Fredon E, Du G, Gerardin C, et al. Oxidized demethylated lignin as a bio-based adhesive for wood bonding. The Journal of Adhesion. 2021;**97**:873-890. DOI: 10.1080/00218464.2019.1710830

[64] Mathieu Y, Vidal JD, Arribas Martínez L, Abad Fernández N, Iborra S, Corma A. Molecular oxygen lignin depolymerization: An insight into the stability of phenolic monomers. ChemSusChem. 2020;**13**:4743-4758. DOI: 10.1002/cssc.202001295

[65] Kumar A, Biswas B, Saini K, Kumar A, Kumar J, Krishna BB, et al. Effect of hydrogen peroxide on the depolymerization of prot lignin. Industrial Crops and Products. 2020;**150**:112355. DOI: 10.1016/j.indcrop.2020.112355

[66] Gierer J. Chemistry of delignification. Wood Science and Technology. 1986;**20**:1-33. DOI: 10.1007/BF00350692

[67] Cymbaluk NF, Neudoerffer TS. A quantitative gas-liquid chromatographic determination of aromatic aldehydes and acids from nitrobenzene oxidation

of lignin. Journal of Chromatography. A. 1970;**51**:167-174. DOI: 10.1016/S0021-9673(01)96851-7

[68] Deng H, Zhang H, Wu Z, Tian X, Jiang X, Guo W. Rules and mechanism for the oxidation of lignin-based aromatic aldehyde under alkaline wet oxygen. BioResources. 2020;**15**:3487-3503. DOI: 10.15376/biores.15.2.3487-3503

[69] Tang P-L, Hassan O, Maskat MY, Badri K. Production of monomeric aromatic compounds from oil palm empty fruit bunch Fiber lignin by chemical and enzymatic methods. BioMed Research International. 2015;**2015**:1-14. DOI: 10.1155/2015/891539

[70] Deng H, Lin L, Liu S. Catalysis of Cu-doped Co-based perovskite-type oxide in wet oxidation of lignin to produce aromatic aldehydes. Energy & Fuels. 2010;**24**:4797-4802. DOI: 10.1021/ef100768e

[71] Voitl T, Rudolf von Rohr P. Oxidation of lignin using aqueous polyoxometalates in the presence of alcohols. ChemSusChem. 2008;**1**:763-769. DOI: 10.1002/cssc.200800050

[72] Xi X, Pizzi A, Frihart CR, Lorenz L, Gerardin C. Tannin plywood bioadhesives with non-volatile aldehydes generation by specific oxidation of mono- and disaccharides. International Journal of Adhesion and Adhesives. 2020;**98**:102499. DOI: 10.1016/j.ijadhadh.2019.102499

[73] Frihart CR, Pizzi A, Xi X, Lorenz LF. Reactions of soy flour and soy protein by non-volatile aldehydes generation by specific oxidation. Polymers (Basel). 2019;**11**:1478. DOI: 10.3390/polym11091478

[74] Yang X, Liang Y, Zhao X, Song Y, Hu L, Wang X, et al. Au/CNTs catalyst for highly selective hydrodeoxygenation of vanillin at the water/oil interface. RSC Advances. 2014;**4**:31932-31936. DOI: 10.1039/C4RA04692G

[75] Sunil More G, Rajendra Kanchan D, Banerjee A, Srivastava R. Selective catalytic hydrodeoxygenation of vanillin to 2-Methoxy-4-methyl phenol and 4-methyl cyclohexanol over Pd/CuFe2O4 and PdNi/CuFe2O4 catalysts. Chemical Engineering Journal. 2023;**462**:142110. DOI: 10.1016/j.cej.2023.142110

[76] Zhang J, Teo J, Chen X, Asakura H, Tanaka T, Teramura K, et al. A series of NiM (M = Ru, Rh, and Pd) bimetallic catalysts for effective lignin hydrogenolysis in water. ACS Catalysis. 2014;**4**:1574-1583. DOI: 10.1021/cs401199f

[77] Zevallos Torres LA, Lorenci Woiciechowski A, de Andrade Tanobe VO, Karp SG, Guimarães Lorenci LC, Faulds C, et al. Lignin as a potential source of high-added value compounds: A review. Journal of Cleaner Production. 2020;**263**:121499. DOI: 10.1016/j.jclepro.2020.121499

[78] Beneroso D, Monti T, Kostas ET, Robinson J. Microwave pyrolysis of biomass for bio-oil production: Scalable processing concepts. Chemical Engineering Journal. 2017;**316**:481-498. DOI: 10.1016/j.cej.2017.01.130

[79] Tian Y, Zuo W, Ren Z, Chen D. Estimation of a novel method to produce bio-oil from sewage sludge by microwave pyrolysis with the consideration of efficiency and safety. Bioresource Technology. 2011;**102**:2053-2061. DOI: 10.1016/j.biortech.2010.09.082

[80] Xie Q, Chen Z, Zhou Y, Pan T, Duan Y, Yu S, et al. Efficient treatment of oily sludge via fast microwave-assisted pyrolysis, followed by thermal plasma vitrification. Molecules. 2023;**28**:4036. DOI: 10.3390/molecules28104036

[81] Halim SA, Swithenbank J. Simulation study of parameters influencing microwave heating of biomass. Journal of the Energy Institute. 2019;**92**:1191-1212. DOI: 10.1016/j.joei.2018.05.010

[82] Sikarwar VS, Zhao M, Fennell PS, Shah N, Anthony EJ. Progress in biofuel production from gasification. Progress in Energy and Combustion Science. 2017;**61**:189-248. DOI: 10.1016/j.pecs.2017.04.001

[83] Yang X, Igalavithana AD, Oh S-E, Nam H, Zhang M, Wang C-H, et al. Characterization of bioenergy biochar and its utilization for metal/metalloid immobilization in contaminated soil. Science of the Total Environment. 2018;**640-641**:704-713. DOI: 10.1016/j.scitotenv.2018.05.298

[84] Molino A, Chianese S, Musmarra D. Biomass gasification technology: The state of the art overview. Journal of Energy Chemistry. 2016;**25**:10-25. DOI: 10.1016/j.jechem.2015.11.005

[85] Cao Y, Chen SS, Zhang S, Ok YS, Matsagar BM, Wu KC-W, et al. Advances in lignin valorization towards bio-based chemicals and fuels: Lignin biorefinery. Bioresource Technology. 2019;**291**:121878. DOI: 10.1016/j.biortech.2019.121878

[86] Xiang Z, Han W, Deng J, Zhu W, Zhang Y, Wang H. Photocatalytic conversion of lignin into chemicals and fuels. ChemSusChem. 2020;**13**:4199-4213. DOI: 10.1002/cssc.202000601

[87] Parkatzidis K, Truong NP, Matyjaszewski K, Anastasaki A. Photocatalytic ATRP depolymerization: Temporal control at low ppm of catalyst concentration. Journal of the American Chemical Society. 2023;**145**:21146-21151. DOI: 10.1021/jacs.3c05632

[88] Xu J, Zhou Z, Liu M, Wang J, Zhang L. Photocatalytic depolymerization of lignin via oxidizing cleavage of Cα-Cβ bonds in micellar aqueous media. International Journal of Biological Macromolecules. 2023;**245**:125476. DOI: 10.1016/j.ijbiomac.2023.125476

[89] Du X, Zhang H, Sullivan KP, Gogoi P, Deng Y. Electrochemical lignin conversion. ChemSusChem. 2020;**13**:4318-4343. DOI: 10.1002/cssc.202001187

[90] Yang C, Maldonado S, Stephenson CRJ. Electrocatalytic lignin oxidation. ACS Catalysis. 2021;**11**:10104-10114. DOI: 10.1021/acscatal.1c01767

[91] Grainger A, Smith G. The role of low carbon and high carbon materials in carbon neutrality science and carbon economics. Current Opinion in Environment Sustainability. 2021;**49**:164-189. DOI: 10.1016/j.cosust.2021.06.006

[92] May AS, Biddinger EJ. Strategies to control electrochemical hydrogenation and hydrogenolysis of furfural and minimize undesired side reactions. ACS Catalysis. 2020;**10**:3212-3221. DOI: 10.1021/acscatal.9b05531

[93] Wang AQ, Li CZ, Zheng MY, Zhang T. Heterogeneous catalysts for biomass conversion. In: The Role of Green Chemistry in Biomass Processing and Conversion. Hoboken: John Wiley & Sons, Inc; 2012. DOI: 10.1002/9781118449400

[94] Diaz LA, Lister TE, Rae C, Wood ND. Anion exchange membrane Electrolyzers as alternative for upgrading of biomass-derived molecules. ACS Sustainable Chemistry & Engineering. 2018;**6**:8458-8467. DOI: 10.1021/acssuschemeng.8b00650

[95] Andrews E, Lopez-Ruiz JA, Egbert JD, Koh K, Sanyal U, Song M, et al. Performance of base and Noble metals for electrocatalytic hydrogenation of bio-oil-derived oxygenated compounds. ACS Sustainable Chemistry & Engineering. 2020;**8**:4407-4418. DOI: 10.1021/acssuschemeng.9b07041

[96] Tu Q, Parvatker A, Garedew M, Harris C, Eckelman M, Zimmerman JB, et al. Electrocatalysis for chemical and fuel production: Investigating climate change mitigation potential and economic feasibility. Environmental Science & Technology. 2021;**55**:3240-3249. DOI: 10.1021/acs.est.0c07309

[97] Rafiee M, Alherech M, Karlen SD, Stahl SS. Electrochemical aminoxyl-mediated oxidation of primary alcohols in lignin to carboxylic acids: Polymer modification and depolymerization. Journal of the American Chemical Society. 2019;**141**:15266-15276. DOI: 10.1021/jacs.9b07243

[98] Ko M, Pham LTM, Sa YJ, Woo J, Nguyen TVT, Kim JH, et al. Unassisted solar lignin valorisation using a compartmented photo-electro-biochemical cell. Nature Communications. 2019;**10**:5123. DOI: 10.1038/s41467-019-13022-7

[99] Li T, Kasahara T, He J, DettelbachKE,SammisGM,BerlinguetteCP. Photoelectrochemical oxidation of organic substrates in organic media. Nature Communications. 2017;**8**:390. DOI: 10.1038/s41467-017-00420-y

[100] Li S, Li Z-J, Yu H, Sytu MR, Wang Y, Beeri D, et al. Solar-driven lignin oxidation via hydrogen atom transfer with a dye-sensitized TiO_2 photoanode. ACS Energy Letters. 2020;**5**:777-784. DOI: 10.1021/acsenergylett.9b02391

[101] Bailey A, Brooks HM. Electrolytic oxidation of lignin. Journal of the American Chemical Society. 1946;**68**:445-446. DOI: 10.1021/ja01207a029

[102] Parpot P, Bettencourt AP, Carvalho AM, Belgsir EM. Biomass conversion: Attempted electrooxidation of lignin for vanillin production. Journal of Applied Electrochemistry. 2000;**30**:727-731. DOI: 10.1023/A:1004003613883

[103] Tolba R, Tian M, Wen J, Jiang Z-H, Chen A. Electrochemical oxidation of lignin at IrO2-based oxide electrodes. Journal of Electroanalytical Chemistry. 2010;**649**:9-15. DOI: 10.1016/j.jelechem.2009.12.013

[104] Lan C, Fan H, Shang Y, Shen D, Li G. Electrochemically catalyzed conversion of cornstalk lignin to aromatic compounds: An integrated process of anodic oxidation of a Pb/PbO_2 electrode and hydrogenation of a nickel cathode in sodium hydroxide solution. Sustainable Energy & Fuels. 2020;**4**:1828-1836. DOI: 10.1039/C9SE00942F

[105] Liu M, Wen Y, Qi J, Zhang S, Li G. Fine chemicals prepared by bamboo lignin degradation through electrocatalytic redox between Cu cathode and Pb/PbO_2 anode in alkali solution. ChemistrySelect. 2017;**2**:4956-4962. DOI: 10.1002/slct.201700881

[106] Cai P, Fan H, Cao S, Qi J, Zhang S, Li G. Electrochemical conversion of corn Stover lignin to biomass-based chemicals between Cu/Ni Mo Co cathode and Pb/PbO_2 anode in alkali solution. Electrochimica Acta. 2018;**264**:128-139. DOI: 10.1016/j.electacta.2018.01.111

[107] Jia Y, Wen Y, Han X, Qi J, Liu Z, Zhang S, et al. Electrocatalytic degradation of rice straw lignin in alkaline solution through oxidation on a $Ti/SnO_2-Sb_2O_3/\alpha\text{-}PbO_2/\beta\text{-}PbO_2$ anode and reduction on an iron or tin doped

titanium cathode. Catalysis Science & Technology. 2018;**8**:4665-4677. DOI: 10.1039/C8CY00307F

[108] Ežerskis Z, Jusys Z. Electropolymerization of chlorinated phenols on a Pt electrode in alkaline solution part I: A cyclic voltammetry study. Journal of Applied Electrochemistry. 2001;**31**:1117-1124. DOI: 10.1023/A:1012280216273

[109] Bruno F, Pham MC, Dubois JE. Polaromicrotribometric study of polyphenylene oxide film formation on metal electrodes by electrolysis of disubstituted phenols. Electrochimica Acta. 1977;**22**:451-457. DOI: 10.1016/0013-4686(77)85100-1

[110] Kumar A, Chandra R. Ligninolytic enzymes and its mechanisms for degradation of lignocellulosic waste in environment. Heliyon. 2020;**6**:e03170. DOI: 10.1016/j.heliyon.2020.e03170

[111] Weng C, Peng X, Han Y. Depolymerization and conversion of lignin to value-added bioproducts by microbial and enzymatic catalysis. Biotechnology for Biofuels. 2021;**14**:84. DOI: 10.1186/s13068-021-01934-w

[112] de Gonzalo G, Colpa DI, Habib MHM, Fraaije MW. Bacterial enzymes involved in lignin degradation. Journal of Biotechnology. 2016;**236**:110-119. DOI: 10.1016/j.jbiotec.2016.08.011

[113] Dashora K, Gattupalli M, Tripathi GD, Javed Z, Singh S, Tuohy M, et al. Fungal assisted valorisation of polymeric lignin: Mechanism, enzymes and perspectives. Catalysts. 2023;**13**:149. DOI: 10.3390/catal13010149

[114] Zhang Y-HP. Production of biofuels and biochemicals by in vitro synthetic biosystems: Opportunities and challenges. Biotechnology Advances. 2015;**33**:1467-1483. DOI: 10.1016/j.biotechadv.2014.10.009

[115] Zeng G-M, Zhao M-H, Huang D-L, Lai C, Huang C, Wei Z, et al. Purification and biochemical characterization of two extracellular peroxidases from Phanerochaete chrysosporium responsible for lignin biodegradation. International Biodeterioration & Biodegradation. 2013;**85**:166-172. DOI: 10.1016/j.ibiod.2013.07.005

[116] Singh D, Zeng J, Laskar DD, Deobald L, Hiscox WC, Chen S. Investigation of wheat straw biodegradation by Phanerochaete chrysosporium. Biomass and Bioenergy. 2011;**35**:1030-1040. DOI: 10.1016/j.biombioe.2010.11.021

[117] Koncsag CI, Eastwood D, Collis AEC, Coles SR, Clark AJ, Kirwan K, et al. Extracting valuable compounds from straw degraded by Pleurotus ostreatus. Resources, Conservation and Recycling. 2012;**59**:14-22. DOI: 10.1016/j.resconrec.2011.04.007

[118] Wong DWS. Structure and action mechanism of ligninolytic enzymes. Applied Biochemistry and Biotechnology. 2009;**157**:174-209. DOI: 10.1007/s12010-008-8279-z

[119] Salvachúa D, Katahira R, Cleveland NS, Khanna P, Resch MG, Black BA, et al. Lignin depolymerization by fungal secretomes and a microbial sink. Green Chemistry. 2016;**18**:6046-6062. DOI: 10.1039/C6GC01531J

[120] Bugg TD, Ahmad M, Hardiman EM, Singh R. The emerging role for bacteria in lignin degradation and bio-product formation. Current Opinion in Biotechnology. 2011;**22**:394-400. DOI: 10.1016/j.copbio.2010.10.009

[121] Antai SP, Crawford DL. Degradation of softwood, hardwood, and grass

lignocelluloses by two *streptomyces* strains. Applied and Environmental Microbiology. 1981;**42**:378-380. DOI: 10.1128/aem.42.2.378-380.1981

[122] Mycroft Z, Gomis M, Mines P, Law P, Bugg TDH. Biocatalytic conversion of lignin to aromatic dicarboxylic acids in Rhodococcus jostii RHA1 by re-routing aromatic degradation pathways. Green Chemistry. 2015;**17**:4974-4979. DOI: 10.1039/C5GC01347J

[123] Sainsbury PD, Hardiman EM, Ahmad M, Otani H, Seghezzi N, Eltis LD, et al. Breaking down lignin to high-value chemicals: The conversion of lignocellulose to vanillin in a gene deletion mutant of *Rhodococcus jostii* RHA1. ACS Chemical Biology. 2013;**8**:2151-2156. DOI: 10.1021/cb400505a

[124] Khan MU, Ahring BK. Lignin degradation under anaerobic digestion: Influence of lignin modifications – A review. Biomass and Bioenergy. 2019;**128**:105325. DOI: 10.1016/j.biombioe.2019.105325

[125] Kato S, Chino K, Kamimura N, Masai E, Yumoto I, Kamagata Y. Methanogenic degradation of lignin-derived monoaromatic compounds by microbial enrichments from rice paddy field soil. Scientific Reports. 2015;**5**:14295. DOI: 10.1038/srep14295

[126] Duan J, Huo X, Du WJ, Liang JD, Wang DQ, Yang SC. Biodegradation of Kraft lignin by a newly isolated anaerobic bacterial strain, *Acetoanaerobium* sp. WJDL-Y2. Letters in Applied Microbiology. 2016;**62**:55-62. DOI: 10.1111/lam.12508

[127] Hu M, Wang J, Gao Q, Bao J. Converting lignin derived phenolic aldehydes into microbial lipid by Trichosporon cutaneum. Journal of Biotechnology. 2018;**281**:81-86. DOI: 10.1016/j.jbiotec.2018.06.341

[128] Yaguchi A, Robinson A, Mihealsick E, Blenner M. Metabolism of aromatics by Trichosporon oleaginosus while remaining oleaginous. Microbial Cell Factories. 2017;**16**:206. DOI: 10.1186/s12934-017-0820-8

[129] Sànchez i Nogué V, Black BA, Kruger JS, Singer CA, Ramirez KJ, Reed ML, et al. Integrated diesel production from lignocellulosic sugars *via* oleaginous yeast. Green Chemistry. 2018;**20**:4349-4365. DOI: 10.1039/C8GC01905C

Chapter 4

Low Ecological Impact Lignin-Based Flame Retardants for Polyurethane Foams

Stanley Olivier Kanemoto, Judith Emery Ngomoyogoli Kanemoto and Madurai Suguna Lakshmi

Abstract

The flammability of polyurethane is a great safety hazard, threatening both lives and goods. Recognizing this, efforts to enhance the fire resistance of polyurethanes can be pursued through various routes. Depending on the classes and applications of polyurethanes, fire retardation can be achieved by incorporating flame retardants or modifying the polymer structure. In response to growing environmental concerns, lignin is an abundant and renewable resource, which has been employed to develop effective flame-retardant polyurethanes, with a simultaneous focus on reducing their ecological impact. Lignin, characterized by its aromatic and phenolic structure, naturally can act as a reactive fire retardant for polyurethanes. Nevertheless, diverse chemical modifications of lignin have been explored to further enhance its fire resistance. This review highlights advancements in the design of phosphorus- and/or nitrogen-containing lignin-based reactive flame retardants tailored for bio-based polyurethanes.

Keywords: polyurethanes, lignin, bio-based flame retardant, chemical modifications, sustainability

1. Introduction

Polyurethanes (PU) are polymers that consist of aliphatic and aromatic portions linked through carbamates in their backbone. The predominant method for producing PU involves the reaction of polyols with a diisocyanate. PU materials are utilized in various forms such as elastomers, coatings, adhesives, and foams, with extensive applications in construction, clothing, medical devices, and electronic equipment [1]. However, PU materials in their original state are generally highly flammable and easily ignited in the presence of a heat source and oxygen [2]. This susceptibility can lead to severe material and equipment damage, compromising their utility in fields where high flame retardancy and thermal resistance are essential [3]. The fire hazards associated with PU have prompted the introduction of legislation and safety standards concerning flammability, and extensive research has been conducted on flame retardants for PU materials [4].

IntechOpen

The flammability of PU poses a significant safety hazard, endangering lives and property [5]. In light of this, the fire resistance of PU can be enhanced through various methods. Depending on the classes and application fields of polymers, fire retardation can be achieved using flame retardants or by altering the polymer structure. Currently, the primary methods for flame retardancy of PU include blending and addition [6], nanocomposites [7], and reaction grafting [8]. Many flame-retardant additive molecules, such as halogen-based compounds, have been banned due to their toxic effects and environmental unfriendliness [9, 10]. These halogen-based flame retardant compounds release toxic gases upon incorporation into the polymer matrix, posing risks to both human health and the environment when the materials burn. Consequently, flame retardants based on phosphorus, nitrogen, silicon, and boron elements have gradually replaced halogenated flame retardants [11].

Among environmentally friendly flame retardants (**Figure 1**), those derived from nitrogen and phosphorus have garnered much attention in industrial and academic communities [12, 13]. Some studies have reported that both fire performance and mechanical properties can be improved by incorporating halogen-free flame retardant nanoparticles, such as graphene [14], graphite nanoplatelets [15], boron nitride [16], and MXenes [17], into the PU system. The effectiveness of these additives is attributed to their higher surface area, reactivity, and better bonding with the PU matrix. With the growing emphasis on environmental and sustainability concerns, there is a push to design safer biobased flame retardants from natural compounds such as lignin and its derivatives, presenting a valuable challenge in PU manufacturing [18].

Lignin, a multifunctional natural polymer, is a component of biomass isolated from hemicellulose and cellulose through industrial pulping processes. For a long time, it has been considered a waste by-product of the paper industry and biorefineries [19]. Lignin possesses a three-dimensional network structure and has become an excellent sustainable chemical raw material due to its biodegradability, renewability, environmental friendliness, and low cost [20].

The utilization of lignin as a biobased flame retardant for PU materials is typically achieved by adding it to polymeric systems through physical blending. This method consists of incorporating lignin into the PU resin at any stage of production and processing. The aromatic structure of lignin imparts impressive properties as a char promoter agent, which acts in the condensed phase and reduces the combustion rate in fire scenarios [21]. However, the addition of a small amount could significantly influence the mechanical properties and the nature of PU materials [20, 21]. The presence of various functional groups, including carboxylic acid, carbonyl, methoxy, and hydroxyl groups [22], makes lignin attractive for incorporating phosphorous and/

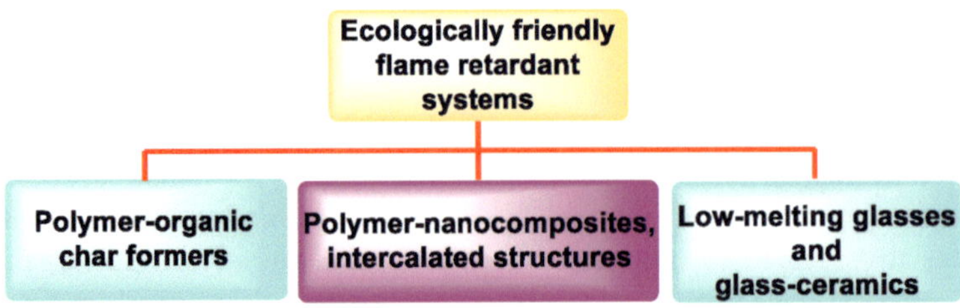

Figure 1.
Classification of ecologically friendly flame retardant systems.

or nitrogen functionalities through chemical functionalization methods [23, 24]. This chapter provides a comprehensive review of the chemical modifications of lignin for effective halogen-free flame retardants, aiming to highlight advancements in the design of phosphorus- and/or nitrogen-containing lignin-based reactive flame retardants tailored for biobased PUs.

2. Lignin-based reactive flame-retardants

Lignin can be valorized as a biobased flame retardant with or without preliminary structural chemical modifications. However, preliminary chemical modifications assist lignin in incorporating more active sites that undergo additional reactions to produce value-added biobased polymers. For instance, chemical, biological, and thermo-catalytic routes are capable of achieving the demethylation of lignin (**Figure 2**) by cleaving aryl-ether linkages [25, 26]. The natural methoxyl groups are replaced with hydroxyl groups in the lignin macromolecule, rendering it more hydroxylated and reactive with a lower molecular mass distribution. Hydroxyl groups in lignin play a pivotal role in the solubility and reactivity of lignin [27]. Additionally, they are easily transformable into new chemical functional groups, expanding the potential of lignin in the continuous production of bio-based polyurethanes [25, 28].

Reactive flame retardancy, also known as intrinsic flame retardancy, can be achieved through various approaches. These approaches primarily involve modifying existing lignin macromolecules by chemically grafting active functional segments onto the lignin backbone or adjusting the structure with potentially active fire retardant functions before the foaming process. The flame retardancy of PU via lignin-based reactive flame retardants involves introducing lignin-containing flame retardant elements, such as phosphorus or nitrogen, into the polyurethane molecular chain through chemical bonding [29]. The development of lignin-based flame retardants presents a sustainable pathway to broaden the application of flame-retardant PU.

One advantage of reactive flame retardants is their ability to be permanently bonded to the main chain. In fact, a relatively low amount of lignin-based reactive flame retardants can achieve a similar effect to that achieved with relatively high loads of additive flame retardants [3]. Conversely, when the preparation of lignin-based reactive systems is not achieved in situ, it could be more expensive and time-consuming. This is because it requires the development of a novel polymer with specific chemical and physical properties.

Based on the differences in flame retardant elements, sustainable lignin-based flame retardant polyurethane can be categorized into phosphorus-containing, nitrogen-containing, and phosphorus- and nitrogen-containing flame retardant PU.

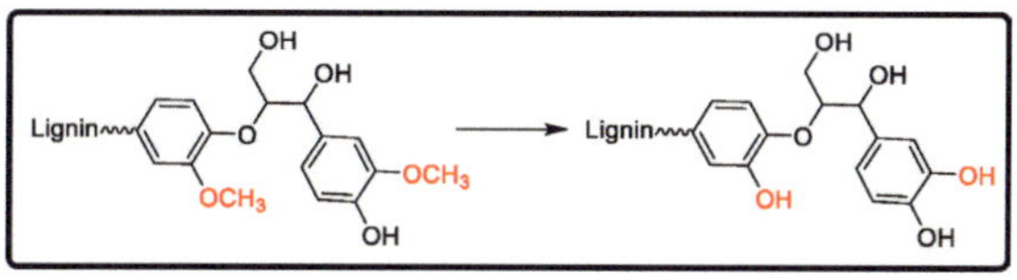

Figure 2.
Scheme illustration of lignin demethylation.

3. Lignin-based phosphorus-containing flame retardants

3.1 Phosphorus-containing flame retardants reactivity

Phosphorus-based flame retardants are the most widely used alternatives to halogen-based flame retardants [9, 12]. Their reactivity in the condensed phase leads to the production of phosphoric acid during the thermal dissociation and decomposition process of the flame retardants. This process, through the rejection of water, produces phosphate compounds. The release of water dilutes the combustible gases. Additionally, they can further act in the condensed phase by catalyzing char formation [10]. This solid residue protects the PU by isolating the non-burned material from the heat, oxygen, and flames. It hinders the volatile products from reaching the flame and feeding it. Phosphorus can take on many oxidation states, as observed in phosphine oxides, phosphines, phosphonium compounds, phosphonates, phosphates, phosphites, and elemental red phosphorus, demonstrating the diverse modes of action of lignin-based phosphorus-containing flame retardants. There is no doubt that the best flame retardants work in more than one way and in more than one phase.

Phosphorus-containing flame retardants from lignin can form volatile active radicals in the gas phase, such as PO2*, PO*, and HPO*, which readily scavenge hydroxyl radicals (OH*) or hydrogen radicals (H*) from the gas phase to inhibit combustion. The mechanism of radical scavenging by phosphorus can be summarized in Eqs. (1)–(3) [30]. The most abundant phosphorus radicals and compounds in the flame are HPO2*, HPO*, PO2, and PO.

$$HPO_2^{*} + H^{*} \rightleftarrows PO + H_2O \quad (1)$$

$$HPO_2^{*} + H^{*} \rightleftarrows PO_2 + H_2 \quad (2)$$

$$HPO_2^{*} + HO^{*} \rightleftarrows PO_2 + H_2O \quad (3)$$

3.2 Functionalized lignin as phosphorus-containing flame retardants

The phosphorylation reaction of aliphatic and phenolic hydroxyl groups is a common method for introducing phosphorus into lignin (**Figure 3**). This process enhances the thermal stability of lignin and facilitates the production of renewable flame retardants for PUs [31]. The thermal decomposition of phosphorylated lignin yields a high residual char through the rearrangement of unstable radicals. The phosphorylation mechanism was elucidated by Qin et al., who investigated the structural

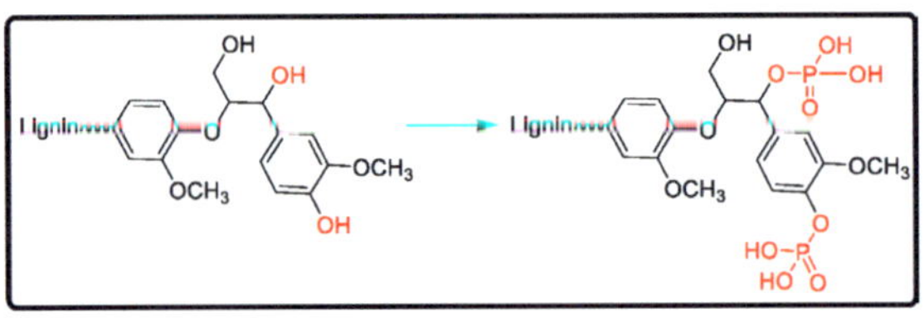

Figure 3.
Scheme illustration of lignin phosphorylation.

sensitivity of softwood lignin during phosphoric acid treatment [32]. They demonstrated that α-O-4 and β-O-4 linkages are prone to breaking during the phosphoric acid–acetone pretreatment process. Phosphorylation of α-O-4 and β-O-4 occurs through a two-step reaction mechanism during the acid treatment. Furthermore, Basso et al. enhanced the hydrophobicity of lignin by employing the phosphorylation reaction on desulfurized kraft lignin using triethyl phosphate at elevated temperatures (90–220°C) [23]. The products resulting from the phosphorylation reaction served as excellent raw materials for the production of lignin-based heat-resistant thermosetting resins.

In a separate study, Gao et al. systematically synthesized phosphorylated kraft lignin using an eco-friendly process that utilized ammonium dihydrogen phosphate ($NH_4H_2PO_4$) as a green and sustainable phosphating reagent [31]. The reactive aliphatic hydroxyl and phenolic groups of lignin underwent a nucleophilic substitution reaction with the phosphate group. The introduction of phosphorus groups into lignin molecules increased the maximum decomposition temperature to 620°C. Prieur et al. developed a straightforward synthetic route for lignin phosphorylation [33], achieving the phosphorylation process using phosphorus pentoxide (P2O5) as the phosphating agent. Phosphate groups were covalently bonded to the complex structure of lignin, promoting dehydration and decarboxylation reactions. Consequently, this significantly increased the amount of residue char.

Zhang et al. devised a pathway for incorporating phosphorus into lignin macromolecules. They synthesized flame retardants containing phosphorus through the chemical reaction of lignin, hexamethylene diisocyanate (HDI), and 9,10-dihydro-9-oxa-10-phosphaphenanthrene-10-oxide (DOPO) [34]. HDI serves as a bridge to connect lignin with DOPO, as illustrated in **Figure 4**. The resulting flame retardant, based on lignin, features a more O=C−P=O conjugated structure, contributing to enhanced flame retardancy and thermal stability. Thermogravimetric analyses (TGA) revealed that the functionalized lignin acts as a proficient char-forming agent, increasing the residual carbon content to 16.55% at 600°C. When these lignin-based flame retardants are employed in the production of flame-retardant PU, the limiting oxygen index (LOI) value of the polyurethane reaches 30.2% with the addition of 15% phosphorus-containing lignin.

In a similar study, Wang et al. synthesized a phosphorus-containing lignin-based flame retardant through in-situ reactions involving lignin, diphenylmethane diisocyanate, and DOPO [29]. The PU produced in a two-step process exhibited excellent flame-retardant properties, evidenced by a gradually increasing LOI value that reached 29.9% with a phosphorus-containing lignin content of 30%. Furthermore, the residual carbon content significantly increased with the phosphorus-containing lignin, reaching up to 34.9% at 700°C.

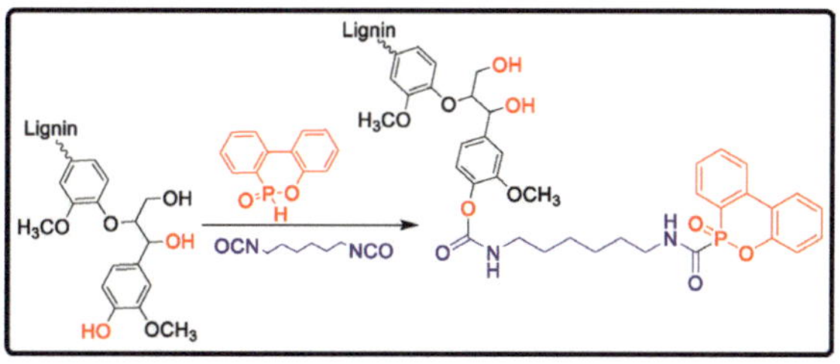

Figure 4.
Schematic illustration of lignin-HDI-DOPO synthesis.

4. Lignin-based nitrogen-containing flame retardants

4.1 Nitrogen-containing flame retardants reactivity

Nitrogen-containing flame retardants encompass a diverse array of condensed-phase and vapor-phase mechanisms that determine flame retardant reactivity. This reactivity is contingent upon both the flame retardant structure and the polymer chain. When PU materials undergo combustion, various species capable of reacting with atmospheric oxygen are formed. Consequently, the propagation of combustion primarily occurs through the reactions outlined in Eqs. (4)-(6).

Nitrogen-containing flame retardants act chemically in the gas phase to scavenge free radicals generated as a byproduct of the thermo-oxidative decomposition of polymer. However, the reaction linked to Eq. (6) is the primary exothermic reaction and provides the most energy to sustain the flame. Hindering the chain-branching reactions related in Eqs. (4) and (5) can slow down or completely stop combustion. Some nitrogen-containing flame retardants can also demonstrate a physical mode of action in the gas phase that either generates a significant amount of non-combustible gases, diluting the flammable gases, or undergo endothermic dissociation, reducing the flame temperature [35]. For instance, nitrogen-containing flame retardants endothermically decompose into incombustible gases, such as nitrogen oxide or ammonia, without smoke. They act by providing both a heat sink and large amounts of inert diluents for the burnable gases in the flame.

$$H^* + O_2 \rightleftarrows HO^* + O^* \tag{4}$$

$$O^* + H_2 \rightleftarrows HO^* + H^* \tag{5}$$

$$HO^* + CO \rightleftarrows CO_2 + H^* \tag{6}$$

4.2 Functionalized lignin as nitrogen-containing flame retardants

Elemental analysis reported in the research work of Ge et al. showed that nitrogen is not an abundant element in the composition of lignin [36]. The most common chemical modifications used to incorporate nitrogen into lignin include nitration and amination.

4.2.1 Nitration of lignin

The nitration reaction of lignin is a straightforward chemical modification process typically conducted in non-aqueous solvents utilizing nitrating agents (see **Figure 5**).

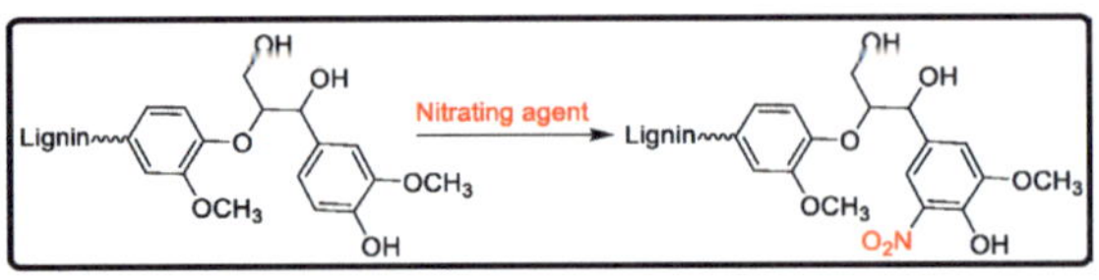

Figure 5.
Scheme illustration of lignin nitration.

Well-established nitrating agents encompass nitric acid in concentrated acetic acid or sulfuric acid [37]. The resulting nitrated lignin is recognized as an amorphous powder with a color spectrum spanning from yellow to brown. Nitration results in the incorporation of approximately 7 wt% of nitrogen into the lignin structure [38].

Veshnyakov et al. optimized lignin nitration using nitric acid in binary mixtures of water and aprotic solvents, such as acetonitrile, dimethyl sulfoxide, 1,4-dioxane, tetrahydrofuran, and dimethylformamide. They employed a nitrating agent formed by mixing acetic acid and the anhydride of nitric acid to produce nitrolignin with a nitrogen content of 4.7% [37]. In another study, Zhang and Huang investigated the thermal stability of lignin-based PU by incorporating nitrolignin into an interconnected network, forming a star-like PU from castor oil and diisocyanate [38]. They observed a slight improvement in the properties of the nitrolignin-incorporated systems. Additionally, nitrolignin was employed as additives or chain extenders after the emulsification stage for the preparation of waterborne PU materials [39]. The reaction between isocyanate groups and hydroxyl groups in nitrolignin produced a crosslinked network in the polyurethane, resulting in a strength of 71.3 MPa.

4.2.2 Amination of lignin

Amination of lignin has been investigated through the Mannich reaction (see **Figure 6**), typically conducted using diethylamine and formaldehyde. The amination of lignin can be integrated into PU synthesis as a reactive precursor or filler to enhance the properties of the composites. Zhou et al. synthesized polyhydric aminated lignin through hydroxylation and amination [40]. The resulting lignin possessed primary amine groups, with a nitrogen content ranging from 0.64% to 4.58%. Polyurethane foams (PUF) prepared from aminated lignin polyol exhibited improved thermal stability and performance compared to those derived from PU using native soda lignin.

On the other hand, Huo et al. generated PUF by blending aminated lignin-based polyol, polyethylene glycol, and diphenylmethane diisocyanates in the presence of water as the blowing agent [41].

The utilization of amines to incorporate nitrogen into the lignin chain presents an intriguing avenue for enhancing the reactivity of lignin. Upon the introduction of amino groups into the lignin structure, its water solubility experiences a direct enhancement facilitated by improved hydrogen bonding and reactivity with isocyanate [42]. Furthermore, aminated lignin can serve as intermediates during the functionalization of lignin. In acidic environments, amino groups readily ionize, acquiring a positive charge, thereby rendering lignin more reactive in aqueous solutions [43].

Amino groups possess the capability to convert hydrophobic kraft lignin into a highly hydrophilic polymer, thereby augmenting certain properties such as

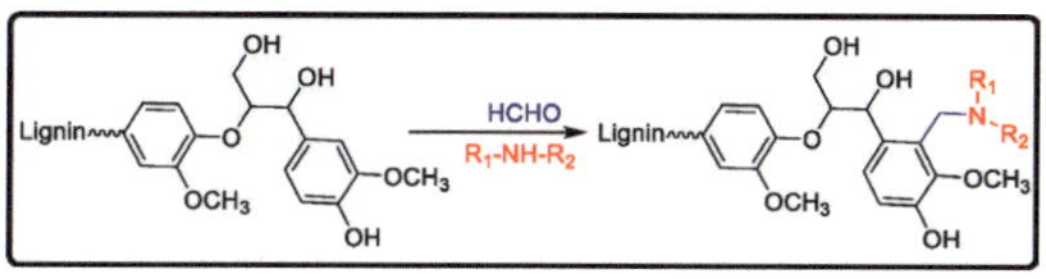

Figure 6.
Scheme illustration of lignin amination through Mannich reaction.

foamability, emulsifying properties, aging resistance, flame-retardant properties, and mechanical strength [44].

5. Lignin-based phosphorus-and nitrogen-containing flame retardants

Most high-efficiency lignin-based flame retardant can be prepared by integrating multiple elements, such as nitrogen, phosphorus, and silicon, in a single flame retardant system [45]. Nitrogen-based flame retardants are very often used as synergists for phosphorus-based flame retardants. They also form an effective class of flame retardants but are treated separately. Some studies have reported that PUs prepared from phosphorus- and nitrogen-containing lignin-based flame retardants offer the most benefits for protecting the environment, as well as improving thermal stability and mechanical properties [9, 13].

Zhu et al. incorporated both phosphorus- and nitrogen-containing functional groups into lignin using a three-stage process, including liquefaction, esterification, and salt formation. Lignin-based flame retardants were prepared through the chemical grafting of phosphate groups and melamine. The lignin-based phosphate-melamine product was then employed to prepare a lignin-based flame retardant for the modification of PUF [46]. After copolymerization between the remaining hydroxyl groups of functionalized lignin and isocyanate, the resulting lignin-modified PUF exhibited excellent flame retardancy in both gas and condensed phases through the escalation of smoke suppression and the increment of thermal properties. The flame-retardant PUF had a LOI of 26.7% and a V-1 rating, as measured by the UL-94 vertical burning test. The presence of rigid aromatic rings in the lignin structure and the covalent bonds between modified lignin and the PU matrix are responsible for self-extinguishment, char residue formation, and inhibition from melt-dripping.

He et al. developed another process by combining mild grafting reactions with the ultrasound method to produce lignin nanoparticles containing phosphorus and nitrogen moieties [21]. The aim was to incorporate functionalized lignin nanoparticles into the formulation of PU so that they could act as both flame retardants and crosslinking agents. Good dispersibility and compatibility in the PU matrix lead to strong tensile properties. The LOI of the resulting PU reached a value of 29.8%, and it received a V-0 rating by adding 10 wt% of lignin-based phosphorus- and nitrogen-containing nanoparticles. Moreover, cone calorimetry tests on PU materials showed that the addition of nanolignin significantly reduced some important parameters during combustion, such as heat release (HR) rate, total HR, and total smoke production rate. The char residues of nanolignin-incorporated PU exhibit high oxidation resistance. The quenching and barrier effects of the char layer improved the flame retardancy performance of lignin-based PU.

Lu et al. investigated the flame retardancy of rigid polyurethane foams (RIPUF) from the lignosulfonate/ammonium polyphosphate intumescent flame-retardant system [47]. The preparation of lignosulfonate-based RIPUF was carried out by copolymerization between lignosulfonate macropolyols and isocyanate. Foams containing 15 wt% of lignosulfonate gave rise to the best thermal stability. They exhibited a lower HR rate (21 $kW.m^{-2}$) and a total HR value of 13 $MJ.m^{-2}$. The incorporation of lignin/ammonium polyphosphate postponed the time-to-peak carbon monoxide production by 96 s compared with virgin RIPUF. The highest LOI over 30% was attained with a lignin/ammonium polyphosphate ratio of 1:5. In addition, lignosulfonate-based RIPUF showed a low-effect heat combustion value, less smoke, and carbon monoxide production.

6. Conclusion

The flammability of polyurethane (PU) materials poses a significant safety hazard, necessitating effective flame retardancy strategies. Current methods for enhancing the fire resistance of PU include blending and addition, nanocomposites, and reaction grafting. The transition from halogenated to environmentally friendly flame retardants, particularly those derived from nitrogen and phosphorus, reflects a growing emphasis on sustainability. This review has focused on lignin, a by-product of the paper industry and biorefineries, as a promising biobased flame retardant for PU materials. Aromatic structure of lignin makes it an effective char promoter agent, reducing combustion rates in fire scenarios. The incorporation of lignin into PU can be achieved through physical blending or chemical modifications. Chemical modifications, such as demethylation, phosphorylation, and amination, enhance lignin reactivity and its ability to serve as an effective flame retardant. The use of phosphorus-containing lignin-based flame retardants involves introducing phosphate compounds that act in the condensed phase, catalyzing char formation and isolating the PU from heat and flames. Nitrogen-containing lignin-based flame retardants, on the other hand, demonstrate diverse mechanisms, including scavenging free radicals and endothermic decomposition, effectively hindering combustion. Combining phosphorus and nitrogen elements in lignin-based flame retardants provides synergistic benefits, enhancing flame retardancy while also improving thermal stability and mechanical properties. Various studies have demonstrated successful synthesis and application of lignin-based flame retardants, showcasing their potential in PU manufacturing. The development of sustainable and environmentally friendly flame retardants aligns with the global shift toward green technologies. As research in this field progresses, lignin-based flame retardants tailored for biobased PU present a promising avenue for addressing fire safety concerns while promoting sustainability in material design and production.

Acknowledgements

The authors acknowledge Prof. Ndikontar Maurice Kor and Prof. Cheumani Yona Arnaud from the University of Yaoundé I, Faculty of Science, Research Unit for Macromolecular Chemistry, for their assistance and support.

Abbreviations

DOPO	9,10-dihydro-9-oxa-10-phosphaphenanthrene-10-oxide ()
HDI	Hexamethylene diisocyanate
LOI	Limiting oxygen index
PUF	Polyurethane foams
PUs	Polyurethanes
RIPUF	Rigid polyurethane foam
TGA	Thermogravimetric analyses
UL-94	Standard established by Underwriters Laboratories
HR	Heat release

Author details

Stanley Olivier Kanemoto[1*], Judith Emery Ngomoyogoli Kanemoto[1] and Madurai Suguna Lakshmi[2]

1 Faculty of Science, University of Yaoundé I, Yaoundé, Cameroon

2 Polymer Science and Technology Division, CSIR-Central Leather Research Institute, Chennai, India

*Address all correspondence to: stanleykanemot@ymail.com

References

[1] Das A, Mahanwar P. A brief discussion on advances in polyurethane applications. Advanced Industrial and Engineering Polymer Research. 2020;**3**(3):93-101

[2] Xu W, Li J, Liu F, Jiang Y, Li Z, Li L. Study on the thermal decomposition kinetics and flammability performance of a flame-retardant leather. Journal of Thermal Analysis and Calorimetry. 2017;**128**(2):1107-1116

[3] Yadav A, de Souza FM, Dawsey T, Gupta RK. Recent advancements in flame-retardant polyurethane foams: A review. Industrial & Engineering Chemistry Research. 2022;**61**(41):15046-15065

[4] Höhne CC, Hanich R, Kroke E. Intrinsic flame resistance of polyurethane flexible foams: Unexpectedly low flammability without any flame retardant. Fire and Materials. 2018;**42**(4):394-402

[5] Lu S, Feng Y, Zhang P, Hong W, Chen Y, Fan H, et al. Preparation of flame-retardant polyurethane and its applications in the leather industry. Polymers. 2021;**13**:1730

[6] Zuber M, Tabasum S, Hussain R, Khan MB, Bukhari IH. Blends of polyurethane-polymethyl methacrylate/TiO2-based composites. Korean Journal of Chemical Engineering. 2013;**30**(8):1652-1658

[7] Marzec M, Kucinska-Lipka J, Kalaszczynska I, Janik H. Development of polyurethanes for bone repair. Materials Science and Engineering: C. 2017;**80**:736-747

[8] Rao WH, Xu HX, Xu YJ, Qi M, Liao W, Xu S, et al. Persistently flame-retardant flexible polyurethane foams by a novel phosphorus-containing polyol. Chemical Engineering Journal. 2018a;**343**:198-206

[9] Gu L, Ge Z, Huang M, Luo Y. Halogen-free flame-retardant waterborne polyurethane with a novel cyclic structure of phosphorus-nitrogen synergistic flame retardant. Journal of Applied Polymer Science. 2015;**132**(3):41288-41296

[10] Naiker VE, Mestry S, Nirgude T, Gadgeel A, Mhaske ST. Recent developments in phosphorous-containing bio-based flame-retardant (FR) materials for coatings: An attentive review. Journal of Coatings Technology and Research. 2023;**20**(1):113-139

[11] Wang D, Wen P, Wang J, Song L, Hu Y. The effect of defect-rich molybdenum disulfide nanosheets with phosphorus, nitrogen and silicon elements on mechanical, thermal, and fire behaviours of unsaturated polyester composites. Chemical Engineering Journal. 2017;**313**:238-249

[12] Nguyen TMD, Chang S, Condon B, Slopek R. Synthesis of a novel flame retardant containing phosphorus-nitrogen and its comparison for cotton fabric. Fibers and Polymers. 2012;**13**(8):963-970

[13] Yuan Y, Yang H, Yu B, Shi Y, Wang W, Song L, et al. Phosphorus and nitrogen-containing polyols: Synergistic effect on the thermal property and flame retardancy of rigid polyurethane foam composites. Industrial & Engineering Chemistry Research. 2016;**55**(41):10813-10822

[14] Zahid M, Castillo AEDR, Thorat SB, Panda JK, Bonaccorso F,

Athanassiou A. Graphene morphology effect on the gas barrier, mechanical and thermal properties of thermoplastic polyurethane. Composites Science and Technology. 2020;**200**:108461

[15] Albozahid M, Naji HZ, Alobad ZK, Saiani A. Enhanced mechanical, crystallization, and thermal properties of graphene flake-filled polyurethane nanocomposites: The impact of thermal treatment on the resulting microphase-separated structure. Journal of Polymer Research. 2021;**28**:1-16

[16] Bashir A, Maqbool M, Lv R, Usman A, Guo H, Aftab W, et al. Surface modified boron nitride towards enhanced thermal and mechanical performance of thermoplastic polyurethane composite. Composites Part B: Engineering. 2021;**218**:108871

[17] Luo Y, Xie Y, Geng W, Dai G, Sheng X, Xie D, et al. Fabrication of thermoplastic polyurethane with functionalized MXene towards high mechanical strength, flame-retardant, and smoke suppression properties. Journal of Colloid and Interface Science. 2022;**606**:223-235

[18] Mhaske ST, Mestry SU, Borse PY. Lignin and its derivatives: Potential feedstock for renewable flame-retardant polymers. In: Bio-Based Flame-Retardant Technology for Polymeric Materials. Oxford: Elsevier; 2022. pp. 133-159

[19] Shao Y, Sun K, Li Q, Zhang L, Wei T, Gao G, et al. Dealkaline lignin–the waste from the pulp and paper industry as acid catalyst in biorefinery. Bioresource Technology Reports. 2019;**7**:100218

[20] Yu O, Kim KH. Lignin to materials: A focused review on recent novel lignin applications. Applied Sciences. 2020;**10**(13):4626

[21] He T, Chen F, Zhu W, Yan N. Functionalized lignin nanoparticles for producing mechanically strong and tough flame-retardant polyurethane elastomers. International Journal of Biological Macromolecules. 2022;**209**:1339-1351

[22] Faleva AV, Belesov AV, Kozhevnikov AY, Falev DI, Chukhchin DG, Novozhilov EV. Analysis of the functional group composition of the spruce and birch phloem lignin. International Journal of Biological Macromolecules. 2021;**166**:913-922

[23] Basso MC, Pizzi A, Delmotte L, Abdalla S. Analysis of the cross-linking reaction of lignin with triethyl phosphate by MALDI-TOF and ^{13}C NMR. Polymers. 2017;**9**(6):206

[24] Yu X, Yang B, Zhu W, Deng T, Pu Y, Ragauskas A, et al. Towards functionalized lignin and its derivatives for high-value material applications. Industrial Crops and Products. 2023;**200**:116824

[25] Podschun J, Saake B, Lehnen R. Catalytic demethylation of organosolv lignin in aqueous medium using indium triflate under microwave irradiation. Reactive and Functional Polymers. 2017;**119**:82-86

[26] Hu L, Pan H, Zhou Y, Hse CY, Liu C, Zhang B, et al. Chemical groups and structural characterization of lignin via thiol-mediated demethylation. Journal of Wood Chemistry and Technology. 2014;**34**(2):122-134

[27] Klein SE, Alzagameem A, Rumpf J, Korte I, Kreyenschmidt J, Schulze M. Antimicrobial activity of lignin-derived polyurethane coatings prepared from unmodified and demethylated lignins. Coatings. 2019;**9**(8):494

[28] Nikafshar S, Zabihi O, Moradi Y, Ahmadi M, Amiri S, Naebe M. Catalyzed synthesis and characterization of a novel lignin-based curing agent for the curing of high-performance epoxy resin. Polymers. 2017;**9**(7):266

[29] Wang Y, Zhang Y, Liu B, Zhao Q, Qi Y, Wang Y, et al. A novel phosphorus-containing lignin-based flame retardant and its application in polyurethane. Composites Communications. 2020;**21**:100382

[30] Kausar A, Rafique I, Anwar Z, Muhammad B. Recent developments in different types of flame retardants and effect on fire retardancy of epoxy composite. Polymer-Plastics Technology and Engineering. 2016;**55**(14):1512-1535

[31] Gao C, Zhou L, Yao S, Qin C, Fatehi P. Phosphorylated Kraft lignin with improved thermal stability. International Journal of Biological Macromolecules. 2020;**162**:1642-1652

[32] Qin W, Wu L, Zheng Z, Dong C, Yang Y. Lignin hydrolysis and phosphorylation mechanism during phosphoric acid–acetone pretreatment: A DFT study. Molecules. 2014;**19**(12):21335-21349

[33] Prieur B, Meub M, Wittemann M, Klein R, Bellayer S, Fontaine G, et al. Phosphorylation of lignin: Characterization and investigation of the thermal decomposition. RSC Advances. 2017;**7**(27):16866-16877

[34] Zhang YM, Zhao Q, Li L, Yan R, Zhang J, Duan JC, et al. Synthesis of a lignin-based phosphorus-containing flame retardant and its application in polyurethane. RSC Advances. 2018;**8**(56):32252-32261

[35] Salmeia KA, Fage J, Liang S, Gaan S. An overview of mode of action and analytical methods for evaluation of gas phase activities of flame retardants. Polymers. 2015;**7**:504-526

[36] Ge Y, Song Q, Li Z. A Mannich base biosorbent derived from alkaline lignin for lead removal from aqueous solution. Journal of Industrial and Engineering Chemistry. 2015;**23**:228-234

[37] Veshnyakov VA, Lakhmanov DE, Yokubjanov MR, Khabarov YG. Nitration of hydrolysis lignin in water-aprotic solvent mixtures. Lesnoy Zhurnal (Forestry Journal). 2020;**5**:184-192

[38] Zhang L, Huang J. Effects of nitrolignin on mechanical properties of polyurethane–nitrolignin films. Journal of Applied Polymer Science. 2001;**80**(8):1213-1219

[39] Huang J, Zhang L. Effects of NCO/OH molar ratio on structure and properties of graft-interpenetrating polymer networks from polyurethane and nitrolignin. Polymer. 2002;**43**(8):2287-2294

[40] Zhou W, Chen F, Zhang H, Wang J. Preparation of a polyhydric aminated lignin and its use in the preparation of polyurethane film. Journal of Wood Chemistry and Technology. 2017;**37**(5):323-333

[41] Huo SP, Nie MC, Kong ZW, Wu GM, Chen J. Crosslinking kinetics of the formation of lignin-aminated polyol-based polyurethane foam. Journal of Applied Polymer Science. 2012;**125**(1):152-157

[42] Liu J, Liu HF, Deng L, Liao B, Guo QX. Improving aging resistance and mechanical properties of waterborne polyurethanes modified by lignin amines. Journal of Applied Polymer Science. 2013;**130**(3):1736-1742

[43] Wang B, Chen TY, Wang HM, Li HY, Liu CF, Wen JL. Amination of biorefinery technical lignins using Mannich reaction synergy with subcritical ethanol depolymerization. International Journal of Biological Macromolecules. 2018;**107**:426-435

[44] Liu Z, Lu X, An L, Xu C. A novel cationic lignin-amine emulsifier with high performance reinforced via phenolation and Mannich reactions. BioResources. 2016;**11**(3):6438-6451

[45] Zhang P, Fan H, Tian S, Chen Y, Yan J. Synergistic effect of phosphorus–nitrogen and silicon-containing chain extenders on the mechanical properties, flame retardancy and thermal degradation behavior of waterborne polyurethane. RSC Advances. 2016;**6**(76):72409-72422

[46] Zhu H, Peng Z, Chen Y, Li G, Wang L, Tang Y, et al. Preparation and characterization of flame retardant polyurethane foams containing phosphorus–nitrogen-functionalized lignin. RSC Advances. 2014;**4**(98):55271-55279

[47] Lu W, Li Q, Zhang Y, Hirose S, Hatakeyama H, Matsumoto Y, et al. Lignosulfonate/APP IFR and its flame retardancy in lignosulfonate-based rigid polyurethane foams. Journal of Wood Science. 2018;**64**:287-293

Chapter 5

Recent Developments in Advanced Lignin-Based Adhesives and Binders

Antonio Pizzi

Abstract

Recently several developments on the use of lignin and its derivatives as wood adhesive and for other binders have come to the fore in the literature. The novelty of these approaches has been dictated by the well-known low reactivity of lignin with aldehydes in its substitution of phenol in phenol-formaldehyde resins. A brief record of the more recent novelties having being published on the use of lignin in the more traditional field of lignin-phenol-formaldehyde (LPF) adhesive resins is reported. However, this review chapter is more focused on the types of more recent alternative approaches that have been used with encouraging results to go around the obstacle of the low lignin-aldehyde reactivity. Thus, approaches based on lignin demethylation coupled with specific oxidation, pre-glyoxalated lignin dialdehyde starch cross-linking by urea, lignin-based non isocyanate polyurethane (NIPU) adhesives and resins, lignin amine coatings, lignin-triethyl phosphate cross-linking for both wood surface coatings and biobinders for metals/Teflon assemblies, and finally direct wood bonding by lignin esterification by citric acid are described.

Keywords: lignin binders, demethylated lignin, lignin specific oxidation, lignin non-isocyanate polyurethanes, lignin-triethyl phosphate binders, citric acid lignin esterification

1. Introduction

It is for a very long time that researchers have been focusing on using lignin for wood adhesives and other binders. The phenolic oligomeric nature of lignin, its relatively low cost and its great abundance as a byproduct of the pulp and paper industry brings to mind its similarity with phenol and has focused on its use as phenolic resins. Thus, for many decades since the early 1960s the literature is extremely rich of articles describing how to substitute lignin fully or partially for phenol in phenol-formaldehyde (PF) resins for wood panel adhesives. Notwithstanding this strong focus on its use the industrial uses for these applications have been extremely rare. Thus in the 1970s and 1980s lignin has been used for up to 30% substitution of phenol in PF plywood adhesives by a major company in the USA and several plywood factories in Canada [1, 2]. This use was discontinued later as it was realized that while the cost of materials was lowered in Lignin-PF resins an equivalent lengthening of the press time resulted from the lignin substitution of phenol, due to the former lower reactivity with aldehydes. The slower press time allowed only the use of these LPF resins in plywood where swiftness in press time was less necessary, hence less limited by the lignin constraining reactivity. Thus, all was reduced to an economic equation: lower material costs but higher processing costs, with the latter being calculated to be predominant, and lignin substitution of phenol in industry was abandoned. Research work on lignin-phenol-aldehyde combinations did not stop however, without really being able to overcome the fundamental drawback outlined above.

More recently a Scandinavian company has been able to manufacture to good effect a plywood adhesive containing 50% lignin substitution of phenol [3] and to successfully market it. This has caused a concomitant increase of interest in other researcher and some literature increase on alternative methods to achieve such a result [4–7]. The publication on these more recent and different approaches are already many, but this review will report briefly only some of them but not focus particularly on them, as they can be considered as an intermediate rather than a final or more advanced solution to the use of lignin in wood adhesives and other binders. This review will mainly focus instead on approaches where the lignin can be used alone or in a great and determinant majority to qualify the adhesive as a bioadhesive rather than to use the still interesting, but more conservative approach described above for LPF resins.

2. Lignin-phenol-formaldehyde adhesives

On this rather more traditional approach a few innovation of note have been proposed that will be very briefly described here. First the substitution of formaldehyde by a much less toxic aldehyde such as glyoxal, an approach still relatively conservative [4, 5].

Then some research work to determine the influence of lignin modification methods on lignin-phenol-formaldehyde (LPF) adhesive properties [6, 7]. Thus, lignin was modified with four different approaches, namely glyoxalation (G), phenolation (P), ionic liquid treatment (IL) (**Figure 1**), and maleic anhydride grafting (MA) (**Figures 2** and **3**). Phenol in PF resins was then substituted with the lignins so modified to a 50 wt% level to prepare modified LPF adhesives for bonding wood particleboard. Standard methods were used to determine the gel times, solids content, density, and viscosity of the these experimental LPF adhesives and the physical properties of the particleboard bonded with them such as internal bond (IB) strength, bending modulus, dimensional stability and even formaldehyde emission levels. Differential scanning calorimetry demonstrated that the MA-pre-reacted lignin presented the lowest curing temperature of all the other experimental LPF resins with modified lignin and non-modified lignin control. They also presented a faster gel time coupled with a higher solids content and viscosity than the resins differently otherwise modified. Moreover, the particleboards bonded with either maleic anhydride modified or ionic liquid modified lignin LPF resins presented lower formaldehyde emission and a much improved mechanical strength than all the others, control included. No significant difference was apparent between the panels bonded with all the experimental modified LPF adhesives.

These works had the merit to show the mechanism of ionic liquid activation on lignin repeating units (**Figure 1**) as well as the effect of maleic anhydride maleation

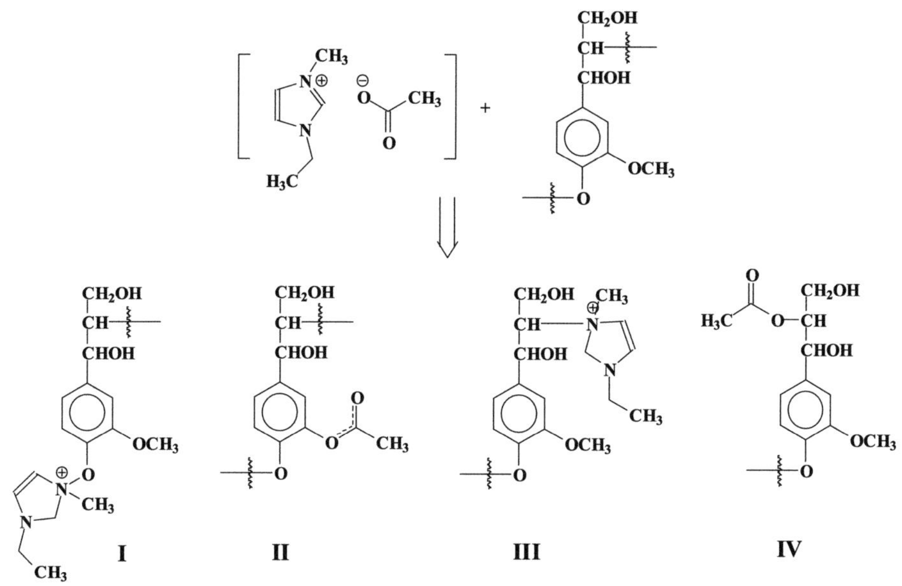

Figure 1.
Action of ionic liquids on lignin units.

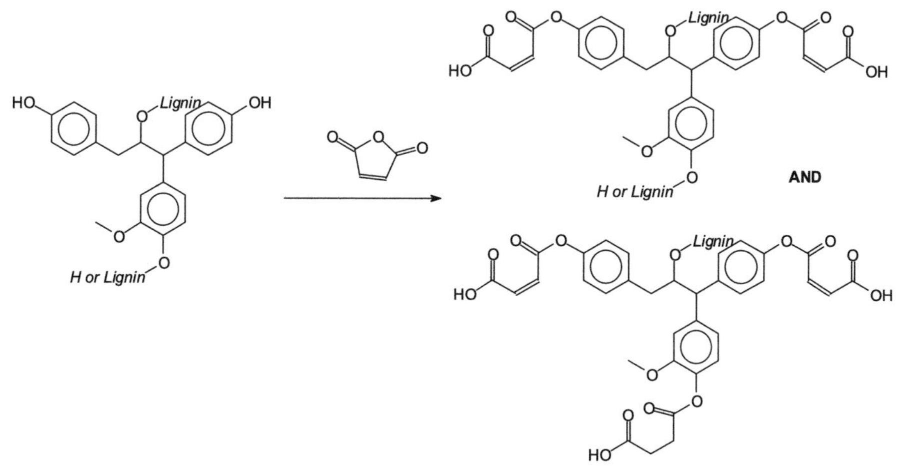

Figure 2.
Reaction of maleic anhydride on lignin units to form maleated lignin.

of lignin increasing the number of reactive site of the material by grafting C=O and C=C linkages on the lignin structure (**Figure 2**) and the way how the maleated lignin linked to the phenol-formaldehyde resin so to form a LPF adhesive (**Figure 3**).

3. Predominant lignin proportions approaches to wood adhesives and binders

The difficulty experienced in developing LPF resins indicates that a different reaction approach has to be taken to avoid the reactivity drawbacks of the material described in the numerous LPF approaches.

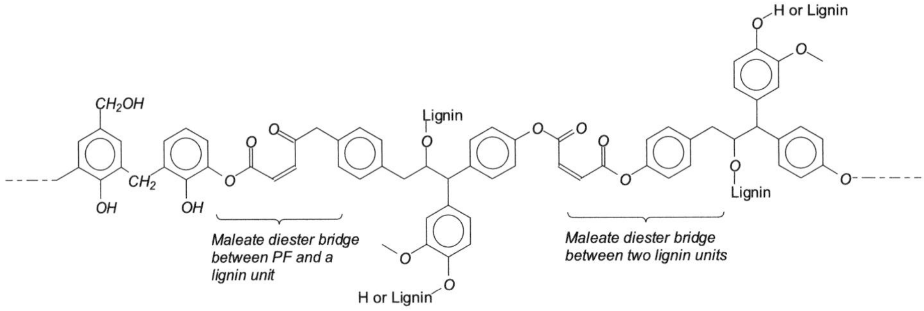

Figure 3.
Maleated lignin bridges with a PF resin.

3.1 Lignin demethylation coupled with specific oxidation

Lignin was successfully modified to prepare a bio-based wood adhesive by just two synthesis steps: demethylation and periodate oxidation [8]. These lignin resins have been based on the reaction of −OH carrying biocompounds such as sugars, protein, tannins etc. to produce vicinal aldehyde groups on the biostructure by cleavage of vicinal C–OH bonds as shown in **Figure 4**.

Specific oxidation of carbohydrates through this route is known for a long time [9] but has recently been applied for all sorts of resins from biomaterials [8–15]. The same is true for lignin. Thus, the basic reaction in the case of lignin develops according to a mechanism that is similar but not identical to that of some other biomaterials (**Figure 5**) [8].

Where in the β-O-4 lignin bond between two lignin units is then cleaved in two sites as shown.

The two aldehyde groups formed still attached to sizable lignin oligomeric residues can than react with the more reactive demethylated groups of the residual lignin to form a hardened network. It must be noted that also other oxidizing agent such as

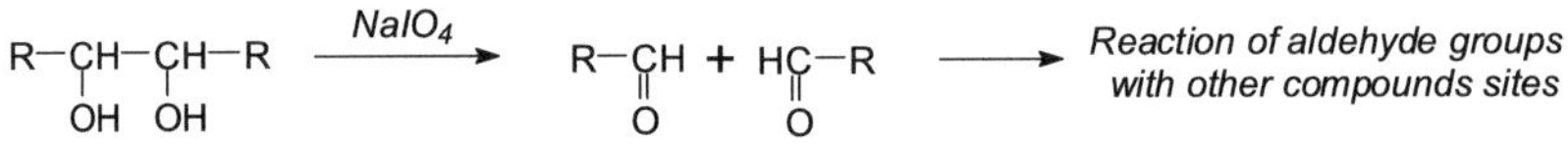

Figure 4.
Schematic example of aldehyde generations by specific oxidation of vicinal C-OH groups.

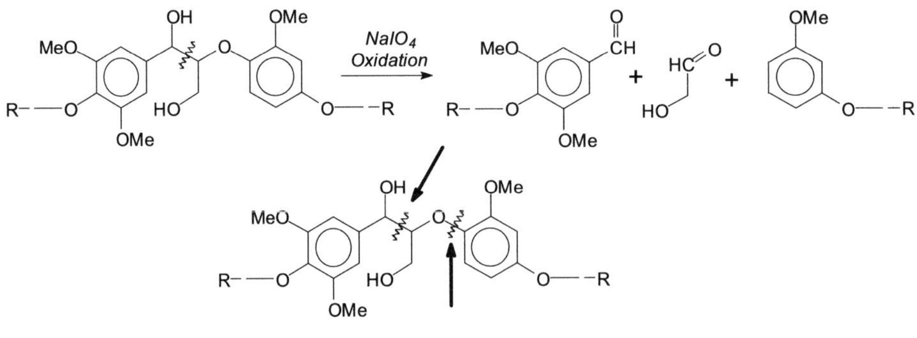

Figure 5.
Lignin cleaving sites and cleaved products by specific oxidation with sodium periodate and other oxidants.

$KMnO_4$ and sodium persulfate can also induce similar cleaving, but only this last has been found to be as specific as the periodate.

Fourier Transform Infrared Spectroscopy (FTIR), Matrix Assisted Laser Desorption Ionization Time of Flight (MALDI-TOF) Mass Spectrometry and Thermomechanical Analysis (TMA) were combined to characterize the bio-based adhesives obtained. Furthermore, the adhesion performance of the bio-based lignin adhesive was measured by testing the tensile shear strength of the bonded joints. The results showed that demethylated lignin gives to the wood joint acceptable dry shear strength, but no wet shear strength. Thus, sodium peroxide ($NaIO_4$) oxidant was used to improve the bonding performance of lignin-based adhesives. The optimal formulation for oxidized demethylated lignin adhesives was found to be by 20% addition of $NaIO_4$. The adhesive presented is a lignin-derived non-aldehyde addition, high biomass content, and its preparation is convenient, low-cost product and can be applied successfully to interior plywood bonding.

In the case of the wet shear strength of oxidized demethylated lignin-based adhesives bonded plywood, nevertheless, is inferior to the requirements of the China national standard (GB/T 9846-2015, ≥0.7 MPa). Therefore, some further modifications will need to be implemented in future research work to reach the relative standard requirements for this aspect.

A further new bio-sourced adhesive approach has been tested based on urea-crosslinked glyoxalated lignin by using dialdehyde starch obtained by hydrogen peroxide oxidation of starch. Urea appears to react with both the dialdehyde starch and the glyoxalated lignin to cross-link the resin to a hardened state [16]. The reaction as determined by instrumental analysis was a shown in **Figure 6**.

The adhesives based on this approach were used to bond wood particleboards, and tested by both differential scanning calorimetry (DSC), and thermomechanical analysis (TMA). The panels internal bond strengths obtained with this approach were satisfying the strength requirements of the relevant European Norms. Cross-linking and bond strength were further improved by adding 5% on resin solids of a bio-sourced glycerol glycidyl ether.

3.2 Lignin-based non-isocyanate polyurethane bioadhesives (NIPU)

The formation of lignin-based NIPU bioadhesives and bioresins is based to a succession of two reactions. First the carbonation of any C–OH carrying compound that can be obtained with a variety of reagents such as cyclic carbonates or even CO_2. However, the system that very recently has succeeds at a relative swift rate of reaction is by reacting C–OH carrying biocompounds with dimethyl carbonate (DMC) to carbonate them. This is a good reagent for the first step to non-isocyanate polyurethanes [17, 18]. DMC is a liquid inexpensive, easily available, that, while flammable, has a pleasant smell similar to methanol and more importantly is classified as presenting no irritating or mutagenic effects, neither by contact nor by inhalation [18].

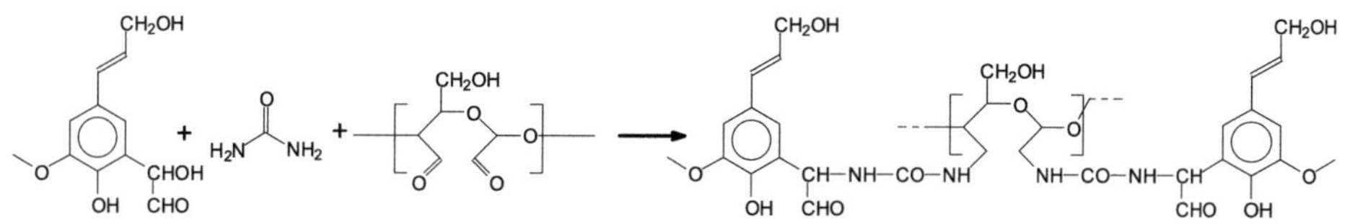

Figure 6.
Reaction of cross-linking by urea of dialdehyde starch and glyoxalated lignin.

DMC-induced hydroxyl groups carboxymethylation is in generally carried out at a temperatures around 90°C, this being close to the DMC boiling point.]It is a bimolecular nucleophilic substitution, acyl-cleaving, under basic catalysis [18]. Recently, some research on preparing polyurethanes without isocyanates from mono and disaccharides [19, 20], or hydrolysable or condensed tannins showed very good performances [17, 21, 22], for wood coating and wood adhesives applications [17, 19–22].

The second reaction is to react the carbonated biomaterial with a diamine. Thus, the schematic sequence of reactions to form a non-isocyanate urethane linkage and NIPU adhesives is as shown in **Figure 7**.

Lignin-based NIPU adhesives for wood particleboards have been developed [23] using organosolv lignin, giving good results. However, good results are obtained only by using hot press temperatures of 220–230°C. While such a hot press temperature is not unusual for modern continuous press wood particleboard factories it is less suitable for older plants. To overcome this drawback a 10–15% addition on total resin solids of an epoxy silane (**Figure 8**) allowed to achieve a very acceptable performance even at the lower temperature of 180°C.

The hardened adhesive crosslink density is increased by the epoxy end of the additive while adhesion to the substrate is rendered easier by the silane end of the additive. MALDI ToF analysis of the oligomers formed during the formation of lignin NIPU resins have shown the formation of compounds with two urethane linkages for a single lignin unit, as well as lignin units dimers linked by a urethane bridge.

The long series of the species formed and found by MALDI ToF leads to a series of linear oligomers that respond to the general structural formula shown in **Figure 9**.

The question of how these NIPU adhesives cross-link to form a hardened tridimensional network should involve a mechanism inducing tridimensional networking, as in reality these NIPU resins cure by just being heated. MALDI analysis detected numerous species presenting tridimensional branching, thus explaining the NIPU resins capacity of cross-linking to a hardened network. The types of oligomers formed can be classed in two categories: (i) oligomers where branching has taken place just by reaction of DMC and the diamine, without reaction with further lignin units, thus branched but without cross-linking, and (ii) branched oligomers where branching has led to cross-linking with several other lignin units in the structure, explaining how these resins can cross-link and harden with heat. An example of this is the compound in **Figure 10**.

Thus, a generalizable formula for the branched type of NIPU resins leading to cross-linking can rather be more correctly shown as in **Figure 11**.

Parasite side reactions do also occur such as for example the reaction of DMC with only the diamine, hence linking two diamines through a DMC molecule, as well as others [23].

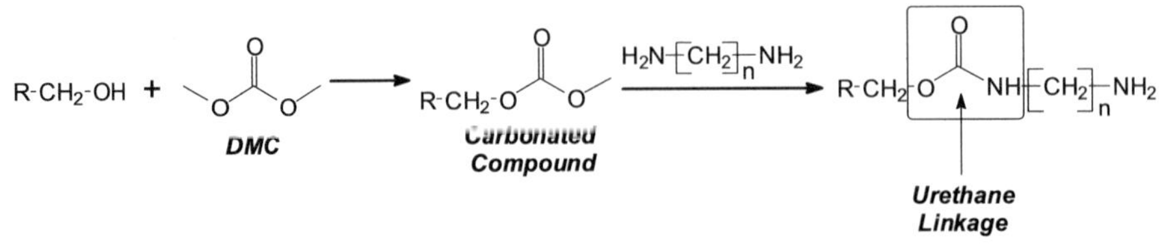

Figure 7.
Sequence of reactions with dimethyl carbonate (DMC) and diamines or polyamines to obtain non-isocyanate polyurethanes (NIPU).

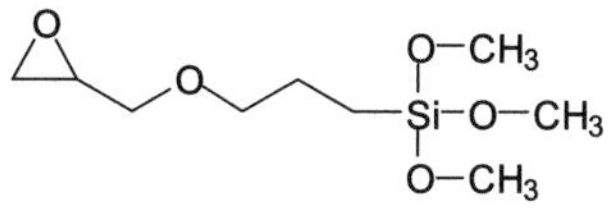

Figure 8.
Silane coupling agent for upgrading lignin NIPU cross-linking [23].

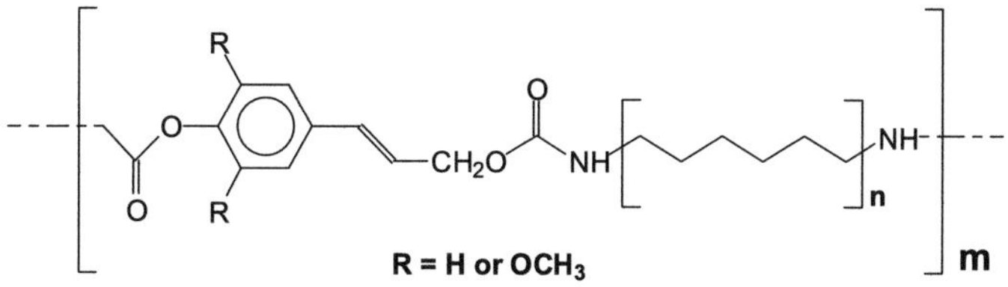

Figure 9.
An example of species formed by reaction of all –OH groups of lignin, both aromatic and aliphatic, with dimethyl carbonate (DMC) and then with the diamine to form urethane linkages [23].

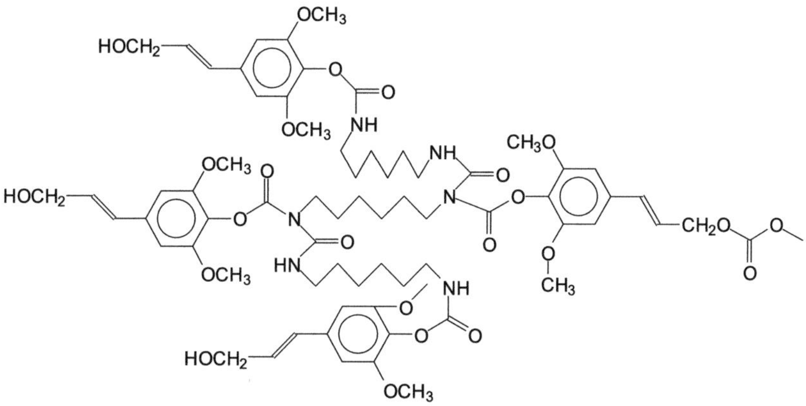

Figure 10.
Example of compounds detected in the formation of lignin NIPU resins showing the cross-linking and networking characteristics of this adhesive system [23].

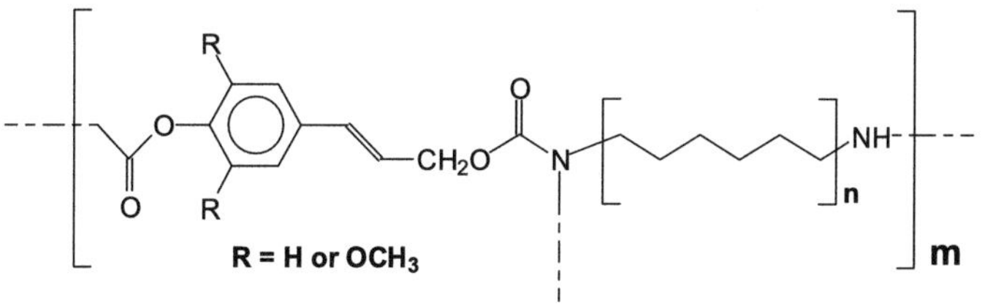

Figure 11.
A schematic representation of a generalized network repeating structure for lignin-based NIPU resins and adhesives.

FTIR spectrometry analysis also confirmed that urethane bridges formed.

A further development of this approach consisted in considerably increasing the bio content of such resins by aminating demethylated lignin to transform any –OH groups present in –NH2 groups by reacting the lignin oligomers mix with ammonia. This reaction has already been used extensively and successfully for other biopolyphenols such

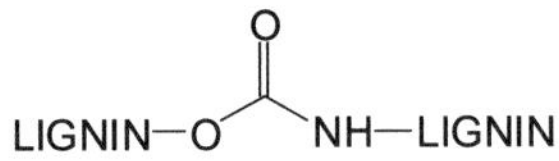

Figure 12.
Structure of lignin NIPU almost 100% of bio green sources.

as condensed tannins [24, 25]. This then allowed the elimination of the hexamethylene diamine, triethylene tetramine and polymamine used to remain with only the –C=O of the DMC residue in the urethane bridges being thus the only one derived from a synthetic compound, but in reality also biosourced as derived from CO_2, thus with a NIPU biocontents of more than 95% to form a structure as shown in **Figure 12**.

3.3 Lignin amine reactions

The limitations to form usable resins that are imposed by the low reactivity of the lignin aromatic rings with aldehydes needed to be overcome by the use of alternative reaction, for example of isocyanates with the lignin methylol groups of hydroxymethylated lignin [26, 27]. This approach, although not "green", yielded the realization that the traditional lignin low reactivity and poor crosslinking can only be overcome for its resins application by using reactions that are not based on the classical phenols-aldehyde approach. While polymeric isocyanates initially served well such a purpose, as they are now much less acceptable for a "green" use due to their uncured toxicity, mainly helped to realize that alternative reaction routes were needed.

It is on the basis of this background that the recent development to crosslink other polyphenols, such as tannins, by alternate polycondensation reactions [27–29] has led to check the possibility of applying these same reactions to obtain new, hardened, cross-linked polycondensation resins based on lignin. One of these approaches was by reacting a commercial, desulfurized kraft lignin with amines [30–32], diamines and with polyamines, to yield hardened resins [33]. In this work (a) covalent bonds were obtained by reacting the amine groups to substitute the phenolic –OH groups on the aromatic rings of lignin. (b) Equally, covalent bonds are also formed by reaction between the amine groups and the aliphatic side chain of lignin units by substituting the lignin side chain alcoholic –OH groups. (c) A proportion of ionic bonds were nonetheless still formed and subsisted between the lignin phenolic and side chain aliphatic –OH groups and the amine at both 180°C and 100°C. (d) Lignin yields hardened resins when reacted with 1,6-hexanediamine at 180°C, and also but to a lesser extent at 100°C.

Precedents to this reaction with amines were also known through reaction of aromatic hydroxyl groups with ammonia [24, 25].

3.4 Lignin triethyl phosphate for wood surface coatings and biobinders for metals/Teflon assembly

A recent research work aimed at obtaining new thermosetting resins by reacting lignin and triethyl phosphate (TEP) for developing through a different set of reactions a range of novel bio-based and heat-resistant paints, lacquers, coatings, resins, and binders for a number of different applications and surfaces [28, 34]. Polycondensation of the two materials appears to occur by reaction of TEP with the phenolic hydroxyl groups of lignin as well as with the aliphatic hydroxyls groups of its side chain. The reaction is favored by higher temperatures and by the presence of ammonia. Initially the reactions was codified by using simple model compounds, namely guaiacol and glycerol for the aromatic and aliphatic parts of lignin. The findings with the two models was then confirmed by reaction of TEP with a desulphurised softwood kraft lignin, namely Biochoice kraft lignin.

The resins prepared by reacting lignin and TEP at 180°C and 220°C produced acetone insoluble rigid, hard, dark solids [34]. Glycerol reacting with TEP at 180°C yielded a liquid transparent resin, but guaiacol reacted with TEP, either with or without NH_3, was so burned that no resin solids were recoverable after being heat treated. Finally, a flexible black semisolid was obtained when the lignin and TEP resin was heated at 90°C, soluble in acetone but not in water [34]. Meanwhile, the TEP heated at 90°C with glycerol, remained a transparent liquid before and after heating it. The resins were evaluated by MALDI ToF, CP-MAS 13C NMR and FTIR spectrometries. The NMR analysis showed, together with some unreacted lignin units, lignin units where TEP had reacted with the lignin phenolic –OH groups, such as shown in **Figure 13**.

The aliphatic –CH2OH of the lignin unit side chain was also shown to react with TEP [34].

The combination of MALDI and 13C NMR brought to the fore a great variety of structures being formed, such as those shown in **Figures 14** and **15**.

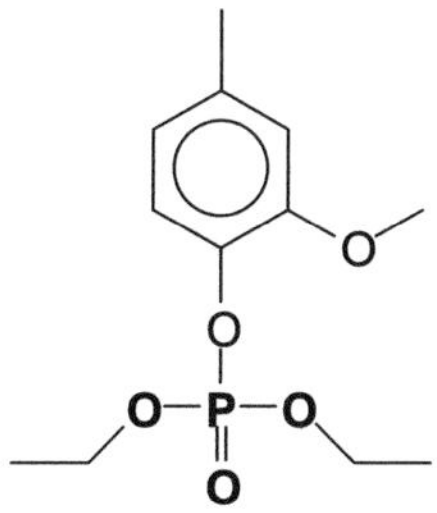

Figure 13.
Mode of linkage of triethyl phosphate with phenolic hydroxyls.

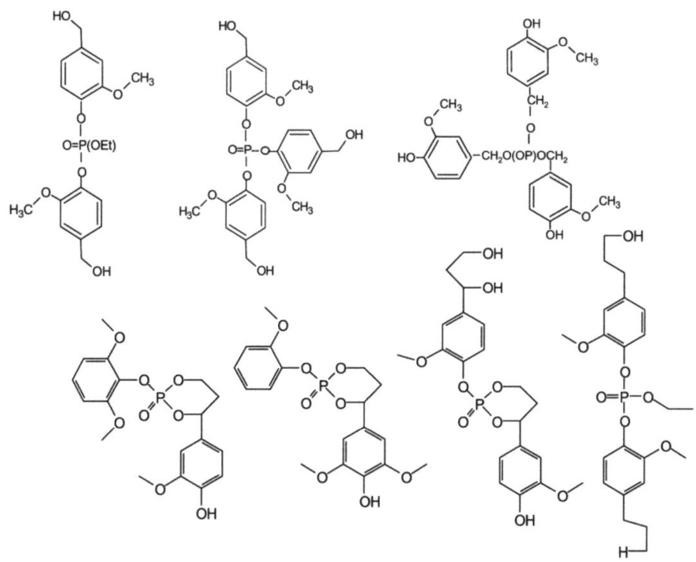

Figure 14.
An example of the variety of structures detected from the reaction of lignin with triethyl phosphate.

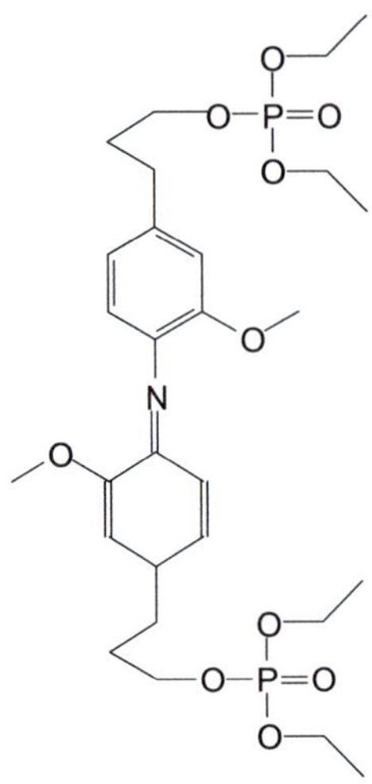

Figure 15.
Example of one of the type of structures formed in the reaction of lignin with triethyl phosphate in which a phenolic hydroxyl has been substituted by nitrogen.

The possibility that the ammonia used in some of the cases has substituted the phenolic hydroxyl groups of two lignin units also exists (**Figure 15**). This type of reaction is facile and has already been reported for other phenolic compounds [24, 25].

This would yield structures of the following type, several of which have effectively been found [34].

For surface coating applications the effectiveness of such a surface coating on wood were determined by a sessile water drop test on the wood surface coated with the lignin and TEP reaction product. The results shown in **Figure 16** do attest the effectiveness of the lignin-TEP coat as a surface finish.

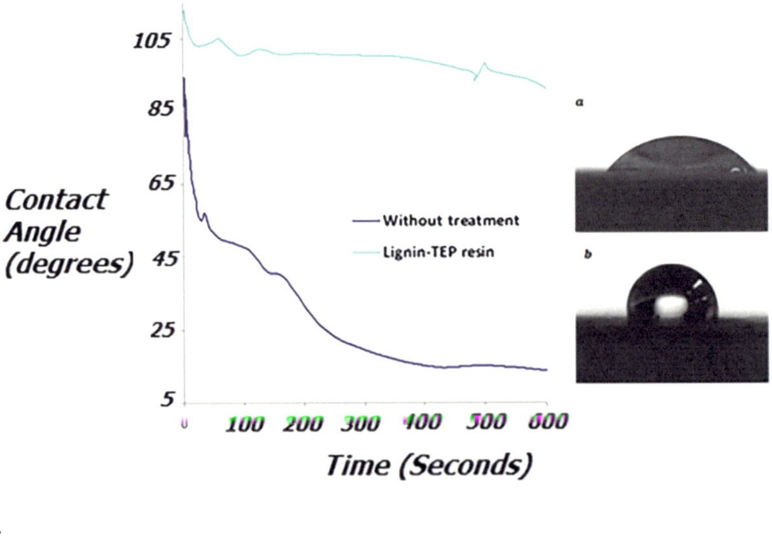

Figure 16.
(left) Water contact angle variation as a function of time of the lignin-TEP-based resin coated beech wood surface and of the untreated beech sample control, and (right) water drop shape after 60 seconds on: (a) an untreated beech wood (control), and (b) a beech wood surface coated with a lignin-TEP-based resin [34].

Figure 16 shows clearly that not only the initial water contact angle is much higher for the surface treated with the lignin-TEP-based resin coating than for the untreated wood surface, but that, moreover, the water contact angle values for the untreated surface decreases rapidly during the testing period, while they remain almost constant for the lignin-TEP resin coated wood surface.

The reaction appears to be dependent on the temperature. Thus, it is favored certainly from 180°C, yielding insoluble hardened resins, as well as when ammonia is present, due to additional cross-linking reactions.

However, the reaction of lignin (and other bio polyphenols, namely condensed flavonoid tannins) with TEP was initially developed for another application, namely as a binder of Teflon on steel or aluminum for non-stick frying pans [28, 34] (**Figure 17**).

The biopolyphenols-TEP binder was developed to substitute phenol-formaldehyde resins. The needs were that it had to bind well both the metal base (steel and aluminum) of the pan and at the same time bind well the surface Teflon coating used for non-stick pans. **Figure 17** shows the case in which the polyphenols-TEP resin was bonded to the metal base without any Teflon (left) and the finished pan with the Teflon applied (right), according to a proprietary process. The test need that was satisfied by this "bio" assembly was that it should be able to repeatedly resist a temperature of more than 400–430°C for more than 11 minutes each time the test was repeated, and it was repeated many times, a test that the 'bio" assembly passed well. Such an application is of course patented for the tannin/TEP system [35], but it is not patented for the equally good lignin/TEP system.

3.5 Citric acid and wood lignin bonding

Citric acid has been promoted as a direct binder for wood [36–38] and also as an additive improving wood binding by traditional or new adhesives [39–42]. While in direct bonding of wood substantial esterification of the polymeric carbohydrate constituents of wood does occur leading to cross-linking between contact surfaces [38], the reaction although less immediately noticed occurs extensively also with lignin [38]. Reaction with model compounds of lignin has led to the identification of citric acid esterified compounds such as shown in **Figure 18**.

As well as esterification of model compounds of carbohydrate constituents of wood such as shown in **Figure 19**.

Figure 17.
Frying pan metal base with the polyphenolic-TEP binder applied to it (left) and finished pan with Teflon applied on the binder according to a proprietary process (right).

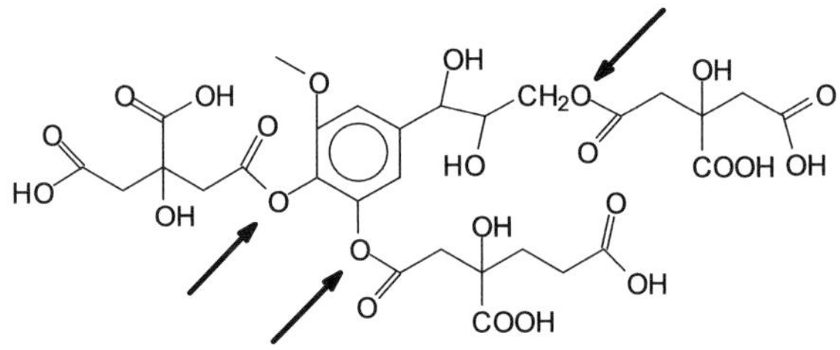

Figure 18.
Example of one of the species detected by reaction of citric acid with a lignin repeating unit [38].

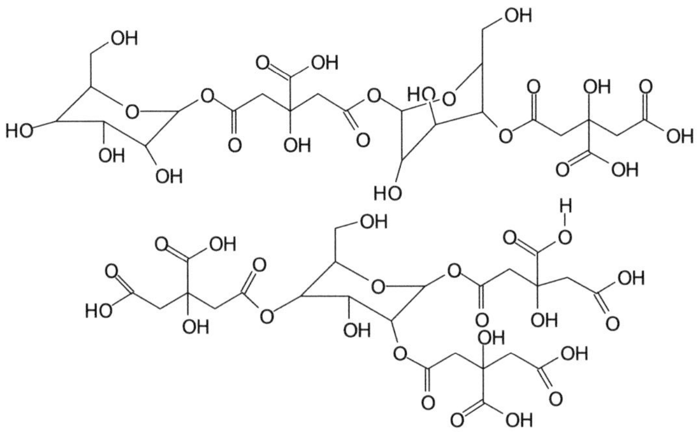

Figure 19.
Example of some of the species detected by reaction of citric acid with glucose as a model compound of wood cellulose and hemicelluloses [38].

Indications are that cross-linking occurs also between polymeric carbohydrates and lignin reinforcing the density of cross-linking of the wood surface network, hence increasing joint strength in plywood and LVL bonding [38]. Some internal rearrangements of lignin also accompany its esterification reaction by citric acid. The bonding results of LVL and plywood panels, thus of flat wood veneers, gave excellent bond strength notwithstanding that the hot pressing conditions used were harsher than what is usual for such an application (180°C for 10 minutes). The bond line interface examined by X-ray densitometry showed that the interface reached peaks up to 1015 kg/m^3 density, this being up to 3 times higher than the density of the veneers between the interfaces (**Figure 20**). Conversely, the density of the veneers zones between the panels bonded interfaces was almost as low than that of veneers not pressed. Such an alternation of high and low densities throughout the panel thickness profiles is similar to what observed for wood welded joints [43].

SEM micrographies showed that the bond line between the veneers even at 3000× magnification (30 μm) was not identifiable, confirming undue densification of the veneers surfaces at the bonded interface [38]. It appeared that the veneers boundary melted and densified under the action of citric acid and that possibly materials from the two veneers surfaces flowed into each other. While the hot-pressing conditions also appeared to also have a marked influence on this effect by contributing to maintain a close contact between the veneers, they were definitely not the major cause of this. This indicated that the reaction between the untreated surface and the citric

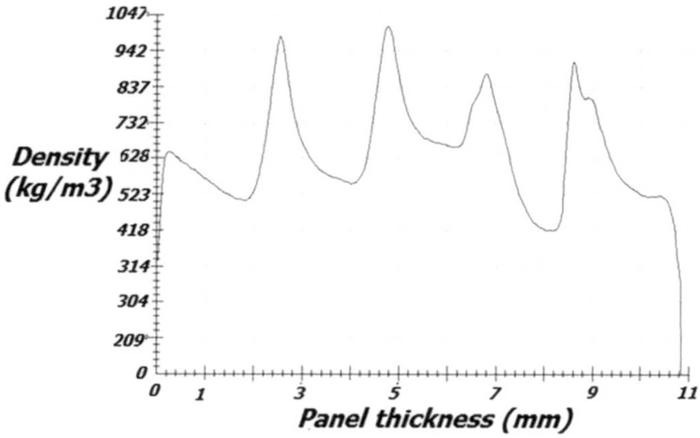

Figure 20.
X-ray density profile of the 5-layer citric acid-bonded LVL [38].

acid treated surface induced a certain level of some wood partial melting causing its flowing, compaction and thus increasing interfacial density.

It is interesting to note that even in wood friction welding the addition of a citric acid solution on the veneers surfaces caused not only increased interfacial melting of the wood substrate but improved water resistance of the welded bond line, indicating the importance of citric acid grafted lignin as regards the improved water resistance of the joint [42].

4. Conclusion

This brief review indicates some alternative routes for the use of lignin in wood adhesives and other binders. It stresses that alternative reactions need to be used if such an abundant natural polymer such as lignin is to take its rightful place in the development of new, more advanced lignin-based "green" materials of all kinds, without the need to reiterate old approaches and ideas on its use. It is also an encouragement for the growing number of applications which will most likely be developed in the future based on lignin and in which lignin will be eventually used. What presented here is only the beginning of what promises to be an even brighter future for lignin applications.

Conflict of interest

The author declares no conflict of interest.

Author details

Antonio Pizzi
LERMAB, University of Lorraine, Nancy, France

*Address all correspondence to: antonio.pizzi@univ-lorraine.fr

References

[1] Terry Sellers Jr, T. Plywood and Adhesives Technology. New York: Dekker, 1985. 661 p. 10.1002/pol.1987.140250208

[2] Calvé L. Lignin-phenol-formaldehyde adhesives. In: Proceedings of the 19th IUFRO Wood Congress; August 1990; Montreal, Quebec, Canada. Vienna, Austria: International Union of Forestry Research Organizations; 1990

[3] Valkonen S, Hübsch C. Lignin based binders: An industrial reality, latest developments. In: Proceedings of the International Conference on Wood Adhesives 2017; 25-27 October 2017. Atlanta, GE, USA: Forest Products Society; 2017. p. 440

[4] Younesi-Kordkheili H, Pizzi A. Improving the properties of phenol-lignin-glyoxal as a wood adhesive by lignin nanoparticles. European Journal of Wood and Wood Products. 2023;**81**:507-512

[5] Younesi-Kordkheili H, Pizzi A. Some physical and mechanical properties of particleboard panels bonded with phenol-lignin-glyoxal resins. Journal of Adhesion. 2020;**96**(16):1385-1395

[6] Younesi-Kordkheili H, Pizzi A. A comparison among lignin modification methods on the properties of lignin-phenol-formaldehyde resin as wood adhesive. Polymers. 2021;**13**:3502. DOI: 10.3390/polym13203502

[7] Younesi-Kordkheili H. Maleated lignin coreaction with phenol-formaldehyde resins for improved wood adhesives performance. International Journal of Adhesion and Adhesives. 2022;**113**:103080. DOI: 10.1016/j.ijadhadh.2021.103080

[8] Chen X, Xi X, Pizzi A, Fredon E, Du G, Gerardin C. Oxidized demethylated lignin as a bio-based adhesive for wood bonding. Journal of Adhesion. 2021;**97**:873-890. DOI: 10.1080/00218464.2019.1710830

[9] Bobbitt JM. Periodate oxidation of carbohydrates. Advances in Carbohydrate Chemistry. 1956;**11**:1-41. DOI: 10.1016/s0096-5332(08)60115-0

[10] Frihart CR, Lorenz L. Specific oxidants improve the wood bonding strength of soy and other plant flours. Journal of Polymer Science Part A: Polymer Chemistry. 2019;**57**:1017-1023. DOI: 10.1002/pola.29357

[11] Xi X, Pizzi A, Frihart CR, Lorenz L, Gerardin C. Tannin plywood bioadhesives with non-volatile aldehydes generation by specific oxidation of mono-and disaccharides. International Journal of Adhesion and Adhesives. 2020;**98**(4):102499. DOI: 10.1016/j.ijadhadh.2019.102499

[12] Frihart CR, Pizzi A, Xi X, Lorenz L. Reactions of soy flour and soy protein by non-volatile aldehydes generation by specific oxidation. Polymer. 2019;**11**(9):1478. DOI: 10.3390/polym11091478

[13] Zhang Q, Song J, Lei H, Pizzi A, Du G, Xi X. Preparation of the water-resistant eco-friendly carbohydrate wood adhesives by Schiff base reaction. Wood Material Science and Engineering. Online 23 August 2023. pp. 1-4. DOI: 1080/17480272.2023.2238261

[14] Guigo N, Mazeau K, Putaux JL, Heux L. Surface modification of cellulose microfibrils by periodate oxidation and subsequent reductive amination with benzylamine: A topochemical study. Cellulose. 2014;**21**:4119-4133. DOI: 10.1007/s10570-014-0459-0

[15] Codou A, Guigo N, Heux L, Sbirrazzuoli N. Partial periodate oxidation and thermal cross-linking for the processing of thermoset all-cellulose composites. Composites Science and Technology. 2015;**117**:54-61. DOI: 10.1016/j.compscitech.2015.05.022

[16] Chen X, Pizzi A, Zhang B, Zhou X, Fredon E, Gerardin C, et al. Particleboard bioadhesive by glyoxalated lignin and oxidized dialdehyde starch cross-linked by urea. Wood Science and Technology. 2022;**56**:63-85. DOI: 10.1007/s00226-021-01344-z

[17] Thébault M, Pizzi A, Dumarçay S, Gerardin P, Fredon E, Delmotte L. Polyurethanes from hydrolysable tannins obtained without using isocyanates. Industrial Crops and Products. 2014;**59**:329-336. DOI: 10.1016/j.indcrop.2014.05.036

[18] Tundo P, Selva M. The chemistry of dimethyl carbonate. Accounts of Chemical Research. 2002;**35**(9):706-716. DOI: 10.1021/ar010076f

[19] Xi X, Wu Z, Pizzi A, Gerardin C, Lei H, Zhang B, et al. Non-isocyanate polyurethane adhesive from sucrose used for particleboard. Wood Science and Technology. 2019;**53**(2):393-405. DOI: 10.1007/s00226-019-01083-2

[20] Xi X, Pizzi A, Delmotte L. Isocyanate-free polyurethane coatings and adhesives from mono- and disaccharides. Polymers. 2018;**10**(4):402. DOI: 10.3390/polym10040402

[21] Thébault M, Pizzi A, Essawy H, Baroum A, van Assche G. Isocyanate free condensed tannin-based polyurethanes. European Polymer Journal. 2015, 2015;**67**:513-526. DOI: 10.1016/j.eurpolymj.2014.10.022

[22] Thébault M, Pizzi A, Santiago-Medina FJ, Al-Marzouki FM, Abdalla S. Isocyanate-free polyurethanes by coreaction of condensed tannins with aminated tannins. Journal of Renewable Materials. 2017;**5**(1):21-29. DOI: 10.7569/JRM.2016.634116

[23] Sarazin J, Pizzi A, Amirou S, Schmidt D, Sernek M. Organosolv lignin non-isocyanate based polyurethanes (NIPU) as wood adhesive. Journal of Renewable Materials. 2021;**9**(5):881-907. DOI: 10.32604/jrm.2021.015047

[24] Braghiroli F, Fierro V, Pizzi A, Rode K, Radke W, Delmotte L, et al. Condensation reaction of flavonoid tannins with ammonia. Industrial Crops and Products. 2013;**44**:330-335. DOI: 10.1016/j.indcrop.2012.11.024

[25] Braghiroli FL, FIERRO V, Izquierdo M-T, Parmentier J, Pizzi A, Delmotte L, et al. High surface-high N-content carbons prepared by hydrothermal treatment of aminated tannin. Industrial Crops and Products. 2015;**66**:282-290. DOI: 10.1016/j.indcrop.2014.11.022

[26] Rahman MM, Zahir MH, Kim HD. Synthesis and properties of waterborne polyurethane (WBPU)/modified lignin amine (MLA) adhesive: A promising adhesive material. Polymers. 2016;**8**:318. DOI: 10.3390/polym8090318

[27] Basso MC, Pizzi A, Lacoste C, Delmotte L, Al-Marzouki FA, Abdalla S, et al. Tannin-furanic-polyurethane foams for industrial continuous plant lines. Polymers. 2014;**6**:2985-3004. DOI: 10.3390/polym6122985

[28] Basso MC, Pizzi A, Polesel-Maris J, Delmotte L, Colin B, Rogaume Y. MALDI-TOF and ^{13}C NMR analysis of the cross-linking reaction of condensed tannins by triethyl phosphate. Industrial Crops and Products. 2017;**95**:621-631. DOI: 10.1016/j.indcrop.2016.11.031

[29] Santiago-Medina F, Pizzi A, Basso MC, Delmotte L, Celzard A. Polycondensation resins by flavonoid tannins reaction with amines. Polymers. 2017;**9**(2):37. DOI: 10.3390/polym9020037

[30] Iranpoor N, Panahi F. Direct nickel-catalyzed amination of phenols via C-O bond activation using 2,4,6-Trichloro-1,3,5-triazine (TCT) as reagent. Advanced Synthesis and Catalysis. 2014;**356**:3067-3073. DOI: 10.1002/adsc.201400460

[31] Yu J, Wang Y, Zhang P, Wu J. Direct amination of phenols under metal-free conditions. Synlett. 2013;**24**:1448-1454. DOI: 10.1055/s-0033-1338703

[32] Hashida K, Makino R, Ohara S. Amination of pyrogallol nucleus of condensed tannins and related polyphenols by ammonia water treatment. Holzforschung. 2009;**63**:319-326. DOI: 10.1515/HF.2009.043

[33] Santiago-Medina FJ, Pizzi A, Basso MC, Delmotte L, Abdalla S. Polycondensation resins by lignin reaction with (poly) amines. Journal of Renewable Materials. 2017;**5**(5):399-399. DOI: 10.7569/JRM.2017.634142

[34] Basso MC, Pizzi A, Delmotte L, Abdalla S. MALDI-TOF and ^{13}C NMR analysis of the cross-linking reaction of lignin by triethyl phosphate. Polymers. 2017;**9**:206. DOI: 10.3390/polym9060206

[35] Polesel-Maris J, Jutang I. Anti-adhesive coatings based on condensed tannins. PCT Patent WO 2017/037393 A1, assigned to SEB S.A. 2017

[36] Umemura K, Ueda T, Munawar S, Kawai S. Application of citric acid as natural adhesive for wood. Journal of Applied Polymer Science. 2012;**123**:1991-1996. DOI: 10.1002/app.34708

[37] Kusumah SS, Umemura K, Oshioka KY, Miyafuji H, Kanajama K. Utilisation of sweet sorghum bagasse and citric acid for manufacturing particleboard I: Effects of predrying treatment and citric acid content on the board properties. Industrial Crops and Products. 2016;**84**:34-42. DOI: 10.1016/j.indcrop.2016.01.042

[38] Del Menezzi C, Amirou S, Pizzi A, Xi X, Delmotte L. Reactions with wood carbohydrates and lignin of citric acid as a bond promoter of wood veneer panels. Polymers. 2018;**10**:833. DOI: 10.3390/polym10080833

[39] Liao R, Xu J, Umemura K. Low Density sugarcane bagasse particleboard bonded with citric acid and sucrose: Effect of board density and additive content. BioResources. 2016;**11**:2174-2185. DOI: 10.15376/biores.11.1.2174-2185

[40] Widyorini R, Nugraha P, Rahman M, Prayitno T. Bonding ability of a new adhesive composed of citric acid-sucrose for particleboard. BioResources. 2016;**11**:4526-4535. DOI: 10.15376/biores.11.2.4526-4535

[41] Del Menezzi C, Pizzi A, Amirou S, Xi X. MUF resins improved by citric acid as adhesives for wood veneer panels. Journal of Renewable Materials. 2023;**11**(2):539-553. DOI: 10.32604/jrm.2022.024971

[42] Amirou S, Pizzi A, Delmotte L. Citric acid as waterproofing additive in butt joints linear wood welding. European Journal of Wood and Wood Products. 2017;**75**:651-654. DOI: 10.1007/s00107-017-1167-x

[43] Leban J-M, Pizzi A, Wieland S, Zanetti M, Properzi M, Pichelin F. X-ray microdensitometry analysis of vibration-welded wood. Journal of Adhesion Science and Technology. 2004;**18**(6):673-685. DOI: 10.1163/156856104839310